Writing for Engineering and Science Students

Writing for Engineering and Science Students is a clear and practical guide for anyone undertaking either academic or technical writing. Drawing on the author's extensive experience of teaching students from different fields and cultures, and designed to be accessible to both international students and native speakers of English, this book:

- Employs analyses of hundreds of articles from engineering and science journals to explore all the distinctive characteristics of a research paper, including organization, length and naming of sections, and location and purpose of citations and graphics;
- Guides the student through university-level writing and beyond, covering lab reports, research proposals, dissertations, poster presentations, industry reports, emails, and job applications;
- Explains what to consider before and after undertaking academic or technical writing, including focusing on differences between genres in goal, audience, and criteria for acceptance and rewriting;
- Features tasks, hints, and tips for teachers and students at the end of each chapter, as well as accompanying eResources offering additional exercises and answer keys.

With metaphors and anecdotes from the author's personal experience, as well as quotes from famous writers to make the text engaging and accessible, this book is essential reading for all students of science and engineering who are taking a course in writing or seeking a resource to aid their writing assignments.

Gerald Rau is Adjunct Professor in the Department of Electrical Engineering at National Chung Cheng University, Taiwan.

Writing for Engineering and Science Students

Staking Your Claim

Gerald Rau

Routledge
Taylor & Francis Group
LONDON AND NEW YORK

First published 2020
by Routledge
2 Park Square, Milton Park, Abingdon, Oxon OX14 4RN

and by Routledge
52 Vanderbilt Avenue, New York, NY 10017

Routledge is an imprint of the Taylor & Francis Group, an informa business

British Library Cataloguing-in-Publication Data
A catalogue record for this book is available from the British Library

Library of Congress Cataloging-in-Publication Data
Names: Rau, Gerald, 1954- author.
Title: Writing for engineering and science students : staking your claim / Gerald Rau.
Description: New York, NY : Routledge, [2019] | Includes bibliographical references and index.
Identifiers: LCCN 2019012283| ISBN 9781138388246 (hardback) | ISBN 9781138388253 (pbk.) | ISBN 9780429425684 (e-book)
Subjects: LCSH: Academic writing—Study and teaching. | Technical writing—Study and teaching. | English language—Rhetoric—Study and teaching.
Classification: LCC P301.5.A27 R38 2019 | DDC 808.06/66—dc23
LC record available at https://lccn.loc.gov/2019012283

ISBN: 978-1-138-38824-6 (hbk)
ISBN: 978-1-138-38825-3 (pbk)
ISBN: 978-0-429-42568-4 (ebk)

Typeset in Bembo
by Swales & Willis Ltd, Exeter, Devon, UK

Visit the e-resources: www.routledge.com/9781138388253

Contents

Figures

Tables

Preface

Every year over 80,000 bachelor's degrees and 50,000 graduate degrees are awarded in engineering in the US alone, and three times that number in science. Worldwide the number is obviously far higher. Because of increasing globalization, many of those graduates must learn to write in English.

Textbooks on scientific and technical writing frequently provide prescriptive models, with templates for different tasks. While these make teaching and grading easy, and students using such texts may excel in class, that does not guarantee success in the real world. Considering the audience and goal will allow a writer to choose which common components of academic writing need to be included in a document, and an understanding of argument structure can guide how to place them. Thus the goal of this text is to provide basic principles that can be applied to any writing task.

The diversity of my training and work experience has played an important role in shaping this textbook. My doctoral degree was in natural science, but I later returned to get a master's degree in science education, so I saw how different writing styles are in the natural and social sciences. While working as a science teacher I wrote lesson plans, lab and classroom activities, various reports, SWOT analyses, and a school accreditation self-study. When a local company was bought by an international corporation, I helped them learn to write email and technical reports to communicate effectively with their new headquarters in the US. My own writing experience has included journal articles, articles for a science teacher magazine, book chapters, and a book. Moreover, my work as a freelance academic editor for over 15 years exposed me to research journal articles from a broad range of disciplines, from linguistics to medicine to applied mathematics, as well as other genres such as grant proposals.

Thus when I was asked to teach academic writing I knew that the structure of articles differed considerably across disciplines, but I believed that engineering used the same basic IMRD (Introduction, Method, Results, Discussion) format as the natural sciences. After all, scientists tend to think of engineering as "applied science." When one engineering student after another said none of their research articles followed that format, I began my own investigation. When my analysis did not match what I had read in textbooks on academic writing, I started to develop my own teaching material.

I have now taught academic writing for seven years, first in the learning center and currently in the electrical engineering department at National Chung Cheng University. My students have come from many branches of engineering, including electrical, mechanical, chemical, and computer science and information engineering, as well as various fields of science and social science. Their analysis of articles in their own fields formed the basis for the general components proposed in this textbook. Combining that knowledge with the principles of argumentation and knowledge construction that underlie modern science education allowed me to generalize and extend the concepts to other types of writing.

This book presents a new general framework for understanding academic and technical writing, particularly in science and engineering, based on the premise that all such writing is argumentation. For academic writing, I claim that although the widely accepted IMRD model is useful in science, it is of limited value for engineering. Support for that claim will be found in the chapters that follow. I further claim that if students learn basic principles of argumentation, learning to think in terms of claim and support, they can become more effective communicators in any genre, written or spoken, as supported by the testimony of former students.

To the student

Many books on science and engineering writing follow a "one-size-fits-all" approach, giving the impression that every field uses the same basic format, and teaching you to write that way. Unfortunately, journal articles in engineering *do not* follow the textbook format for journal articles in science, and there is a lot of variation between different disciplines, and even subfields within a discipline. The same could be said for any other type of writing.

Thus this book approaches writing from a genre analysis perspective, which simply means learning to write like someone *in your field* by analyzing good examples of writing *in your field*. It provides step-by-step instructions through various stages of the writing process, with examples and practice exercises. The book will be most useful if you have a competent mentor to check your work, either an instructor in a writing course or an academic advisor (or preferably both), but can be used for self-study as well.

Since I have lived in Taiwan for most of the last 25 years and most of my students have been native speakers of Chinese or other Asian languages, the text naturally addresses the needs of those whose first language is not English. Additional material for those using English as a lingua franca may be found on the publisher's website. Nevertheless, since no one uses academic English as their first language, any novice writer should be able to benefit from the ideas presented here. I hope you will find the book useful.

To the teacher

There are several new ideas in this book that have never been published before, although some will be appearing in journals and a book chapter at about the same time. The first is the proposal of IPTC (Introduction, Process, Testing, Conclusion) as a way of naming the prototypical format for engineering research articles, based on my own analysis of articles in engineering journals. As I will explain, the focus of an engineering article is on developing and testing a new design to solve a problem, rather than on explanation of data to answer a question. This is reflected in an argument structure and topic placement that differs from the traditional IMRD format of scientific papers. I also propose the 7Cs of Change, which evolved gradually through several years of teaching stepwise revision of academic writing. The chapter on email likewise contains novel ideas related to well-known sociolinguistic and pragmalinguistic principles.

Applied linguists will find that sometimes I do not follow standard terminology in the field. Since I am not a linguist by training, the first draft of the book was written before I realized I was in many cases reinventing the wheel. Even after reading the linguistics literature, I have chosen in some cases to retain my own terminology. Because this book is written for science and engineering students, I have tried to use terminology they will readily understand rather than the fine distinctions teased out by linguists over the years. In the notes you will find the standard linguistic terminology and references to some of the seminal works, in many cases published decades before I stumbled on a similar concept. Where my analysis differs from previously published work, often based on analysis of texts from engineering that do not match previous studies of research articles in science, this is also discussed in the notes, since it is of more interest to teachers than students.

A grant from National Chung Cheng University allowed me to hire research assistants to confirm many of my preliminary findings, but there is still much to be done. Relatively little linguistic analysis has been conducted on the structure of engineering research articles or technical reports, so this text should by no means be considered the final word. Rather, it is my hope that this work will stimulate further research by providing a productive framework to test and modify. Any exercise in this text, conducted on a selection of journal articles from one field or journal, or a corpus from a certain business or industry, would provide a suitable topic for a paper or thesis in applied linguistics. Comments from engineers are also highly welcome, whether to correct misconceptions, make the categories clearer or more useful, or expand the work to other areas. If you have any suggestions on how the book could be improved, please let the publisher know and

I will very happily correct any mistakes, give better examples, and cite your research or comments in any future editions.

This text is too long for a one-semester class. Proposed syllabi and suggestions on how to use the text in basic and advanced classes may be found in the supplemental teaching materials on the publisher's website. On the website you will also find PowerPoint summaries of each chapter and additional exercises and materials to assist those using English as a lingua franca. Teaching tips in each chapter give suggestions on how to make your class more successful, based on my experience. This book works very well with a flipped-classroom approach, where students read the textbook at home and come to class to do the exercises and homework. This allows you to catch problems quickly, resulting in less frustration for the students and less correcting for you.

Finally, note that the book is primarily addressed to students, not teachers. In many cases I will address the reader as "you," referring to the one learning to write. It is my intent to come alongside them, as much as is possible given the medium, acting as a mentor and sharing anecdotes from my own experience rather than providing dry formal pabulum that they would dread reading.

Acknowledgments

First and foremost, my appreciation goes to my wife, D. Victoria Rau, a Professor in the Institute of Linguistics at National Chung Cheng University, who identifies herself as an applied linguist although much of her work is in Austronesian linguistics and sociolinguistics. She pointed me to the books of Swales and Feak and urged me to pursue this work. Without her, it would never have gotten off the ground.

Students in my classes over the years have contributed to my knowledge of academic writing in their fields and pointed out where what I said in class did not match what they saw in their exemplar articles. From the engineers who took my first writing class (when I still thought that engineering followed the IMRD format) to the present, their questions and comments have caused me to question and improve my analysis and clarify both the writing and the exercises.

Work on an early draft of this textbook was partially supported by a teaching material development grant from the Ministry of Education, 2016 Teaching Excellence Program, administered by National Chung Cheng University (榮獲105年度教育部獎勵大學教學卓越計畫及國立中正大學補助). The grant allowed me to hire three graduate assistants, Vithong Nguyen from the Institute of Linguistics and Wei-Young Chen and Jin Tsung-Hsin Liu from the Department of Electrical Engineering, to confirm and refine my preliminary analysis of various aspects of engineering articles. Jennifer Kuo, a very capable undergraduate in linguistics at Dartmouth College, also assisted with the work and contributed important ideas to the book and related articles that will appear. Without them, the task would have been far more difficult.

Christine Feak provided encouragement and helpful comments on an early draft of this work. John Blake was willing to use a draft copy of this book in his class on thesis writing at the Japan Advanced Institute of Science and Technology and offer feedback. Yun-yin Huang, Anita Pu, Ching-Yen Ellie Yu, and Tetyana Smotrova, writing instructors who attended my workshop at ROCTEFL 2018, also provided important encouragement and suggestions. Please accept my thanks.

Francesca McGowan, my editor at Taylor & Francis, was willing to support a proposal from an unknown author, and Elizabeth Cox, her capable editorial assistant, handled many of the details of the process. I am also indebted to the five anonymous reviewers of the project whose comments led to its approval, as well as all the members of the copyediting and production teams. The diagrams and flowcharts in the book were drawn by Romer Valera.

Various works by Swales and Feak provided an introduction to genre analysis, as well as many of the linguistic terms, as will be seen in the numerous citations. The approach taken in this book is very similar to that of Cargill and O'Connor (2013), and I have benefitted from reading their work. Although the first draft was complete before I heard of theirs, during the revision process I have combined some insights from their work with my own observations, and cited places where they had already published similar ideas.

Thanks to various publishers for permission to use previously published graphics, including Oxford University Press (Table 7.2), John Wiley and Sons (Table 24.2), and the Seismological Society of America (Table 24.3). Thanks also to Gregory T. Kleinheinz et al. for permission to reprint the Open Access article "Effects of rainfall on *E. coli* concentrations at Door County, Wisconsin beaches" in the Sample Component Analysis, as well as to InterVarsity Press (US and UK) for permission to quote from my previous book, *Mapping the Origins Debate*, in Class Exercise 21.3 (both in the online supplemental material).

Undoubtedly there are many other works in the field that should have been cited, ideas that I pulled from various sources and used in my teaching before I decided to write for publication. I have tried to track them down, but undoubtedly have missed some. Any oversights are unintentional, and any remaining errors my own.

Many thanks go to my blind masseur at Siloam Helping Hands who saved my neck (and back, and shoulders) multiple times along the way. Finally, special thanks to El Shaddai and the nurses and doctors at Dalin Tzu Chi Hospital, without whom I would not have been able to see this book to completion.

Reference

CARGILL, M. & O'CONNOR, P. 2013. *Writing scientific research articles: Strategy and steps*, New York, John Wiley & Sons.

Part I

Getting the big picture

Chapter 1

General principles of writing

1.1 Fitting in, standing out

I currently teach at a university in a rural area in southern Taiwan. There is no written dress code, but almost everyone dresses informally. Students come to class in shorts and tee shirts, with sneakers or flip-flops on their feet. It just seems to fit the area. Someone wearing high heels and the latest fashion would seem out of place, perhaps a visitor from elsewhere. They just don't fit. But among the informally dressed students, some stand out. They ask the right questions, work hard, and excel in class.

The same principle applies to writing in science or engineering, whether in academics or industry. For example, there may not be a written code that says what a research article should look like in your field, but everyone who has been in it for a while knows the unwritten rules. If you follow those rules, you will fit in. If not, they will look at you like an outsider, and your paper is less likely to be accepted. But within those papers that all look similar, some stand out. They ask the right questions, are well written, and wind up being highly cited. Nevertheless, if you don't fit in, you never get the chance to stand out (Figure 1.1).

The problem is that when you begin to write in a new field of study, you do not know the expectations, the unwritten code. I suggest there are three ways to learn the unwritten expectations in your field:

(1) Gradually learn by experience;
(2) Ask someone with more experience;
(3) Purposefully study the structure of successful writing.

Many students choose the first route, often by default rather than active choice. They finish their research and are ready to write, but then realize they have no idea how to structure their writing. Nevertheless, they try their best, and wind up very discouraged when their teacher or advisor says they have completely missed the mark. After a number of attempts, they finally succeed, and after a few more attempts manage to master the style. Nevertheless, they may not be able to describe it to others because they have never considered what makes some writing better *structurally*, only in terms of content. This is also the limitation of the second route—some professors can train their students to do the research but have never studied or thought critically about how to write.

That brings us to the third route, and the purpose of this book. Every journal has its own unwritten rules that define the expected structure for articles, and you must learn

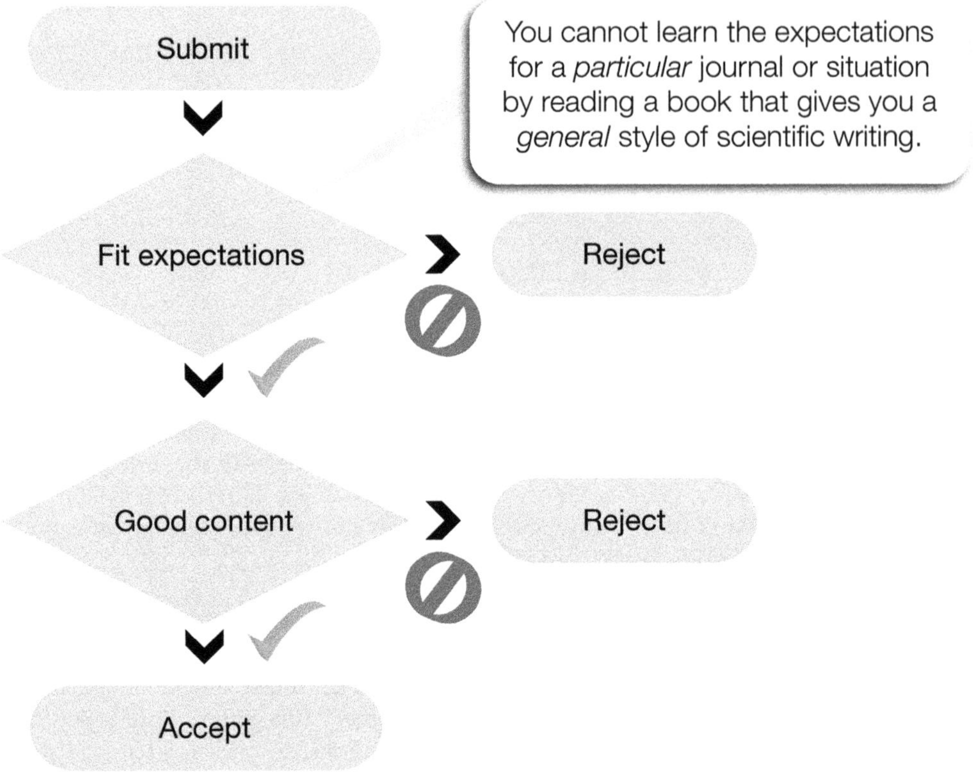

Figure 1.1 Fitting in and standing out.

what those rules are if you want your work to be accepted by that journal. Every business has similar unwritten rules for reports in that company. You cannot learn the expectations for a *particular* situation by reading a book that gives you a *general* style for a certain type of writing. You can do it by reading and working for many years and gradually assimilating the style, or you can expedite the process by actively studying the structure of good examples. In this book you will learn and practice how to do that.

The process initially will seem very tedious. At first you may wonder if it is worthwhile, but it has two advantages. The first and most immediate is that whatever documents you need to write as a student will require less revision and get higher marks,

> The method taught in this book allows you to write better, with less revision, and easily adapt your writing to meet the expectations of different situations.

because they meet the expectations. The second and ultimately more important is that it will be easier to adapt your writing to different situations. After using the method to purposefully study journal articles, you can compare any subsequent writing task with what you already know. By noting the differences, you will be able to adjust your style of writing accordingly. If you have learned to write in one format by trial and error, you will have to use the same slow method for the second as well.

1.2 We have to start somewhere, but why research articles?

No matter what you want to learn, it pays to start with the fundamentals. Therefore, no matter what you plan to write first, we will begin with an analysis of research articles. There are four good reasons for this:

> **Research articles are readily available examples of good arguments; understanding their argument structure will help you read and write better.**

(1) You will learn to *read more effectively* to get the information you need to work in science or engineering;
(2) Research articles are *complete arguments*, thus a good example for your own writing, but also part of an ongoing research effort;
(3) Other genres of student and technical writing are *predictable modifications* of research article structure in that field;
(4) Research articles are *readily available*.

What the American novelist Stephen King said is no less true for academic writing than fiction: "If you don't have time to read, you don't have the time (or the tools) to write. Simple as that" (King, 2000: 147). Whether in school, academics, or industry, you will need to read research articles to get up-to-date information on the latest advances in your field. As we will see, research articles have a relatively predictable order of claims. Articles include most of the components common in science and engineering writing, and certain phrases are frequently used to mark each component. Understanding that structure and recognizing those phrases will help you locate the information you need for your own work more quickly and efficiently.

Although each research article stands alone as a complete argument, it is also part of the growing body of knowledge in that field. Thus each article is one link in a long chain of research and makes sense only in the context of that whole chain. You will need to read many articles before you can see the overall structure of that chain and understand which links are most important in holding the whole chain together, and which merely connect one item dangling from that chain like a charm on a bracelet. The same is true for technical reports, email, or any other type of writing discussed in this book.[1]

Furthermore, as we will see in Chapters 10–17, although other genres make different basic claims than journal articles, they include many of the same components, evidence, and reasoning and are strongly influenced by the research article structure in that field. Thus understanding the structure of journal articles will assist you with almost any type of writing.

From a practical perspective, research articles are easy to obtain. Although it is difficult to access authentic examples of industry reports, grant applications, reference letters, and many other genres because of confidentiality concerns, there is a virtually inexhaustible supply of journal articles available in any field.

1.3 Academic English: a new language

> "Is this English? Why can't I understand it?" Academic English is not anyone's first language.

If you are a graduate student, or even an upper-level undergraduate, you have probably read an academic journal article. What was your first impression? Many students, even those whose first language is English, wonder, "Is this English? Why can't I understand it?" The language used in journals is very different from the language of everyday speech.

Linguists (those who study the structure and use of language) tell us that there are many different genres of writing. Genre refers to a specific category of written, spoken, musical, or artistic presentation, and each genre has its own style. Academic English is different from everyday English, and each discipline even has its own discipline-specific English, with its unique vocabulary and expressions. Technical writing is not the same as academic writing in the same field. The technique we will be using in this book is based on genre analysis. It will help you learn to both read and write academic and technical English, particularly the discipline-specific English of your field.

Since academic English is not a first language for any of us, this book should be useful to any beginning writer. As with learning any other new language, it will take time, but the principles taught in this book will help you improve faster. Eventually you will learn to not only write but also think in the patterns expected in your field.

1.4 Identifying good exemplar articles

> The best exemplar is one that is well written and has a good argument structure, not the one closest to your topic, most cited, or written by the top researcher.

Throughout the book you will be examining exemplar articles. As the name implies, you can use exemplars as examples when you do similar writing, so it is important to identify which documents will be good examples. A good exemplar article is not necessarily the one whose content is the closest to your topic, or the most cited, or one written by the most famous researcher in the field. The most-cited papers often lay out a new theoretical framework or methodology, and therefore may use a different format than a typical research article in the field. Some of the best-known researchers in a field can get away with poor writing because their research is respected, an advantage not shared by new authors. Instead, look for an article that seems to be *well written*, more from the perspective of argument structure than grammar.

The first thing to notice in choosing exemplar articles is the genre. Since our first goal is to analyze standard research articles, *do not use the following as exemplars*:

These are other genres that may provide good information for your research but are not good exemplars because they have a different purpose and structure.

- Review or survey articles
- Overview of a standard[2]
- Brief reports, letters, communications
- Conference papers
- Trade publications
- Popular science publications

Reviews and surveys are usually written by experienced senior researchers. While these may be an excellent place to start reading what has been done in your field, they do not present any new research, and therefore lack many components. The overview of a standard may likewise be vital to your research, but the structure and purpose are very different, with no new testing or data. Usually the title or a text box at the top of the first page will indicate if an article fits in any of these categories.

Brief articles and conference papers are often based on preliminary results, shorter, and not as carefully reviewed, and they generally do not have as developed an argument structure as full articles. Brief articles may also be called letters or communications, and usually have those words in the heading of the first page of the article, or perhaps in the name of the journal. Any article whose reference includes Proceeding, Workshop, or Symposium is probably a conference paper.

Trade and popular science publications are written for people in industry or the general public, respectively, and are usually a summary of applications of the latest research rather than the research results of a single group. They are easily identified by their brighter colors and graphics.

Recent articles generally make the best exemplars, because expectations change over time.

It is also important when deciding which articles to use as exemplars to consider the date of publication. Recent articles are better, because both research and writing styles change over time.

Every journal has slightly different expectations and requirements. You might expect to find differences between, say, chemical engineering and electrical engineering, but there are also huge differences within electrical engineering and from journal to journal within a field (Rau, In preparation). Thus, it is easier to begin by analyzing several articles from one journal. After you are familiar with the process, doing it for another journal will be easier. Eventually you will be able to quickly analyze an article's format based on comparison with those you already know and extend the knowledge to other types of writing.

> Starting with three articles by different authors from one journal allows you to determine the expectations of that journal, so you can fit in.

In the homework for this chapter, you will identify three exemplar articles by different authors in your field. Why three? you may ask. With only one or two it is impossible to establish what is typical of your field, but with three it is likely that at least two will be similar, giving you a better picture of the expectations.

1.5 Learning the fundamentals

> Learning to write is like learning to paint. You learn a particular genre by copying successful work.

Writing is an art. To become an artist, a student must first learn basic elements like shape and color, then principles of composition. After learning the basic techniques, the student may be asked to create their own work, copying the style of a certain master. Similarly, if you want to write like the experts in your field, you have to study their work and try to duplicate the style, but with your own content. That is the fundamental principle underlying the use of exemplars, good examples of a genre. At the same time, as you improve, you will develop your own unique style.

In Chapter 2 you will learn the overall format of academic writing in your field, using research articles as a prototypical example. In Chapter 3 I will introduce the premise of this textbook: that all writing in science and engineering is (or should be) structured as a series of arguments, with nested claims supported by evidence from both past and present work, linked by logical reasoning. Then, in Chapters 4–6 you will examine in detail the claims made in different divisions of research articles. Chapters 7–9 complete the picture, as you learn to identify the evidence and sequencing strategies used to support the claims.

1.6 Exploring different genres

In the next part of the book (Chapters 10–17), I will show how, once you have learned to identify and emulate the format of an argument in a journal article, you can apply the same principles to any writing. Just as it is easier to pick up a new genre of art once you have mastered one, the same is true for writing. The principles of using an exemplar and focusing on argument structure can be applied to any genre of undergraduate, graduate, academic, or technical writing. They can be extended further to job applications, oral presentations, and even email correspondence.

> Once you master the technique, you can apply the same principles to any other writing.

As long as you focus on your audience and the underlying claims that need to be addressed for each type of writing, you will be able to adapt what you have learned to write successfully in any situation. Your first attempts may not be perfect, but should be close enough that others can provide specific suggestions for improvement.

1.7 Creating your own masterpiece

One of the hardest things for new writers to understand is the number of revisions required. Some portions may survive with minimal changes, but those involving integration of work from different authors or justification of a plan of action are much harder to write and will require extensive revision. That being the case, it does not make sense to polish every sentence when you write it the first time, or even to write everything in the order it will appear in the final document.

> One of the hardest things for new writers to understand is how many times you will have to rewrite your work.

Rather than filling your paper from left to right, top to bottom, it makes sense to think of how a painter creates a masterpiece. An experienced painter begins with a broad outline, sketching in the overall size and shape of various components, then chooses one of the main themes, filling in the broad brush strokes of that across the whole canvas, then another, and finally adding all the small details. That is how I am going to encourage you to write.

In Chapter 18 I will describe in more detail the reason for writing in stages. In Chapter 19 you will learn how to prewrite your article, brainstorming about what should go into your document and putting those ideas in order, following your own template derived from exemplars. In Chapter 20 I will give suggestions on keeping track of what you have written, which is particularly essential in academic writing where it may take over a year from the time you start until an article is published.

Although only one chapter is devoted to prewriting and one to writing, the next three chapters (21–23) are devoted to various aspects of rewriting, highlighting its importance in creating your masterpiece. In these chapters you will learn about the 7Cs of Change (Coherence, Conciseness, Connection, Connotation, Consistency, Correctness, Collaboration), seven steps of rewriting that will gradually transform your writing from very rough to a highly polished argument.

In the last part, Chapters 24–27, we will turn our attention to graphics, references, and the title and abstract, important elements of academic writing which are normally finalized after most of the writing and rewriting are finished. Once your work of art is complete, you want it to be noticed and appreciated. To do that, you need to convince someone to put it on display, so in the final chapter we will look at how to submit your work and respond to reviewer comments.

Evaluate your understanding

(1) What are the advantages of the method proposed in this book?
(2) Why will we begin by analyzing the structure of research articles, no matter what you plan to write first?
(3) What types of papers should you *not* use as exemplars, and why?
(4) What is the advantage of using three exemplar articles rather than only one, and three from the same rather than different journals?

Homework

(1) Beginning with one article related to your field, find at least two other articles from the same journal that might serve as exemplars.

Hint: Use Google Scholar or another academic search engine to find recent articles that have cited the first.

Hint: It is preferable to use articles by three different authors, to ensure that the style is that preferred by the field, not a single author.

(2) List the references from your three articles and check that they are all standard research articles.

Hint: If you know how to use a reference manager (section 25.3), do so. If not, type the title in Google Scholar. At the bottom of the listing, click on the quotation mark. Copy one of the formats into a Word document.

Hint: Look for signs of other genres in the reference (Letters, Proceedings . . .).

Teaching tip: Show examples of each type of article and how to recognize them before asking students to submit potential exemplars.

Teaching tip: It is essential to evaluate students' exemplars early, to make sure all are standard research articles, not other genres. It will help students if you can tell them which of the three articles is most prototypical or best written.

Teaching tip: Alternatively, ask students to submit five articles, and help them select three that appear to be better exemplars, rather than finding that some are not suitable and having to resubmit other options.

Teaching tip: Obviously, if all the students in your class are writing another genre and have access to exemplars, you could begin with that instead of research articles, but by doing so they do not learn how to read research articles.

Teaching tip: Sometimes authors will use the same title for a conference paper and a journal article. This can cause confusion when downloading the reference, especially if the article is an early access preprint version, common in engineering. Tell students to check that the reference and header of the article match, and how to reference preprint versions.

Notes

1 Linguists call this intertextuality (the connection between texts). See section 14.1.

2 In engineering, new designs must conform to accepted standards. Overview articles reduce the hundreds of pages of detail in the complete standards to tens of pages of general description.

References

KING, S. 2000. *On writing: A memoir of the craft*, New York, Scribner.

RAU, G. In preparation. IPTC: A prototypical format for engineering research articles.

Chapter 2

Overall format of research articles

2.1 One format or two?

Our first step in learning how to *write* will be learning how to *read* research articles. There are several good reasons for this, explained in section 1.2. Research articles are the heart of science and engineering, reporting the latest advances, so you need to learn to read them. If you find them difficult to read, you are not alone. The vocabulary, grammar, and overall structure are far more difficult than popular writing. Nevertheless, it is easier if you understand how they are structured, and thus where to find the information you need. After learning their structure, I will show you how to use that knowledge to write whatever you want.

The IMRD format (Introduction, Methods, Results, and Discussion, also called IMRaD) is a well-established tool for teaching academic writing. While this structure is easy to identify in many science articles, there are problems applying it to many branches of engineering. In this chapter I propose an analogous format, IPTC (Introduction, Process, Testing, and Conclusion), based on several distinctive features. By the end of the chapter you will know how the formats differ and which is used in your field.

2.2 Research article formats: IMRD vs. IPTC

Research articles in science and engineering exhibit obvious differences in the number, names, and length of sections. Since the format of research articles in science has been better studied and is well established, we will look at those first, then show how typical engineering articles differ, followed by a justification of the need for a new designation for the prototypical format of engineering articles based on systematic differences between science and engineering. In section 2.3 I will discuss how variation in research methodology leads to predictable differences from the prototypical format within a field or sub-field.

2.2.1 IMRD and modified IMRD (IMRaDC) formats

Science research articles usually have four sections and use general section titles. The Results and Discussion comprise the longest and most important part of the article, particularly if tables and figures are included. This is illustrated from two sample articles, which I will refer to by shortened titles rather than authors' names, for ease of recognition.[1]

Throughout this book *I will use the term "section" to refer to the actual named sections of an article, and "division" when referring to the conceptual units of a format.* A division sometimes contains several sections.

> **"Section"—the numbered and named portion of a research article; "Division"—a conceptual unit, including one or more sections.**

The section numbers, titles, and length of each section for an article from a microbiology journal are shown in Table 2.1 (Kleinheinz et al., 2009). For this article, the *section* titles follow the prototypical format for science articles, matching the *division* names taught in almost all textbooks on scientific writing: Introduction, Materials and Methods, Results, and Discussion (IMRD). Subsection titles, on the other hand, are specific to the article. Comparing the length of divisions in this article, we find that in word count the Results division constitutes over half of the article, although the paragraph count is only one third, because of the extensive tables.

> **In IMRD, there are usually four sections, with general titles closely related to the division names. Results and Discussion are the longest divisions.**

This format may be modified, as shown in Table 2.2 for a chemical engineering article (Peng et al., 2017). Here the Results and Discussion are combined into one section, with a separate Conclusion. This modified IMRD format (Introduction, Materials and

Table 2.1 Section titles and length in an IMRD article, "Effects of rainfall"

Div.	*Section/Subsection titles*	*Paragraphs*	*Word count*	*%*
I	1. Introduction	6	787	15
M	2. Materials and Methods			
	2.1. Water Sample Collection	2	257	
	2.2. Water Sample Collection After Rainfall Event	2	449	16
	2.3. Sample Analysis	1	48	
	2.4. Statistical and Graphical Analysis	1	79	
R	3. Results	9	2903	55
D	4. Discussion	6	788	15
Total		27	5311	100

Table 2.2 Section titles and length in an IMRaDC article, "Pt-Ni nanocage"

Div.	*Section/Subsection titles*	*Para-graphs*	*Word count*	*%*
I	1. Introduction	4	513	11
M	2. Experimental section			
	2.1. Chemicals	1	69	
	2.2. Catalyst preparation	1	296	
	2.3. CCM preparation	1	118	24
	2.4. Characterization	1	250	
	2.5. Electrochemical measurements	1	113	
	2.6. CCM testing	1	243	
R&D	3. Results and discussion			
	3.1. Physical characterization	3	773	61
	3.2. Ex-situ RDE characterization	2	395	
	3.3. CCM morphology and PEMFC performance	6	1545	
C	4. Conclusions	1	160	4
Total		22	4475	100

Methods, Results and Discussion, Conclusion) we will call *IMRaDC*.[2] In this article, both the second and third section are divided into numbered and named subsections. Although the name of the second *section* is "Experimental section," a general title common in this journal, it clearly describes the materials and methods used, and would also be referred to as the Materials and Methods *division*.[3] Over half of the article is in the Results and Discussion.

The Introduction division in IMRD situates the current research relative to previous research and states the research goal.

Subdivisions are common in the Materials and Methods division, but specific topics vary by field and research method.

Each division has a specific purpose. In IMRD (or IMRaDC) format the purpose of the Introduction division is to demonstrate the importance of and need for the work and to state the research goal. This places the current research in the context of what has already been done. In a few fields there may be a separate second section for literature review within the Introduction division.

This is followed by the Materials and Methods division, which provides details about how the research was done. This may be divided into subsections to make it easier for readers to find certain information, which in natural science may include materials purchased, material preparation procedures, material characterization, data collection procedures, and data analysis procedures (often including statistics). Specific topics will vary by field and research method.

> Data (evidence) and interpretations (inference) are in separate sections in IMRD, but combined in one section in IMRaDC.

Data summaries are provided in the Results division, often in tables or graphs, sometimes with accompanying statistical test results. Interpretation of that data is found in the Discussion. If the two are combined, there is almost always a separate Conclusion to summarize the main points of the article, often by answering the research questions, if that is how the goal was stated.

2.2.2 IPTC and modified IPTC (IPC) formats

> In IPTC, there are usually more than four sections, many with titles specific to the research. The second division, describing the new design, is the longest.

Unlike science, engineering articles usually have more than four sections, with several section titles specific to the research. Moreover, the new design presented in the second division is the focus. Other differences will appear as we look more closely at each division.

To highlight these and other differences that will be described later, I propose a new format, IPTC, with division names analogous to IMRaDC, as shown in Table 2.3. To justify why this is necessary we must examine how the analogous divisions differ. Since neither the section number nor titles match IMRD, we will first consider how to assign sections to divisions.

> The Introduction division in IPTC may be as short as one paragraph with no citations or several pages long with graphics. It may include more than one section.

There is quite a bit of variation in the first division of IPTC format. Like IMRD, the purpose of the Introduction is to establish the importance of and need for the work and to state the research goal. Almost all engineering articles begin with a section titled Introduction. Some articles also have a second section that contributes to this purpose, just as a literature review would in social science. Other Introductions are a single section, but that section may be over a page long and contain tables and figures, very rare in IMRD. The shortest Introductions may be only one or two paragraphs, with no citations of previous work. Each journal tends to have Introductions with a similar length and structure, but there is great variation between journals.

Table 2.3 Analogous divisions in IMRaDC and IPTC formats

IMRaDC	*IPTC*
Introduction	Introduction
Materials and Methods	Product or Process
Results and Discussion	Testing
Conclusion	Conclusion

> The second division in IPTC, which is usually the longest, describes the new Product or Process in one or more sections with descriptive titles specific to the research.

The second division, the focus of the paper and therefore usually the longest, describes the current research, whether it be development of a new Product or a new Process. The analogy between these names and Materials and Methods should be clear, but most IPTC articles report the development of only one entity, as reflected in the use of the singular and "or" rather than "and." In natural sciences the name of the second division is often shortened to Methods, a convention I will follow throughout this book. Similarly, I will refer to the second division of IPTC as Process.

> The Testing division in IPTC is comparable to a combined Results and Discussion division. It may include several sections.

The third division discusses the testing of the new solution and how it compares with the current best solution or a standard. This is comparable to a combined Results and Discussion division. Although many articles use some variant of the term Experiment, Testing is a better name because the test rarely fits the current preferred definition of an experiment, which involves manipulation of an experimental variable and measuring the effect on a response variable. This is undoubtedly a result of lax use of the term experiment in many science textbooks and laboratory guides to denote any sort of data collection. It may also reflect that the definition of experiment has changed over time (Bazerman, cited in Swales, 1990: 113).

> The one- or two-paragraph Conclusion summarizes the contribution of the article to the field.

A brief Conclusion summarizes the contribution of the article to the field. This corresponds to the Conclusion in IMRaDC or the final paragraph or two of the Discussion in IMRD.

Now we will apply these principles to identifying the divisions of three engineering articles, beginning with one from electrical engineering, shown in Table 2.4 (Sayginer & Rebeiz, 2016). The section numbers and titles are written in the style used by the journal. Section II is assigned to the Introduction because it gives background on the standard architecture rather than details about the current work. Section III, the new method developed and described in the article, comprises almost half of the total paper, whereas the corresponding Method division accounted for less than 25% of the IMRD articles. Sections IV and V describe how the new design is tested and the results of those tests, respectively.

Next, in Table 2.5, we will look at an article from computer science and information engineering (Hu et al., 2013). Again, there is a second section in the Introduction that describes Related Work, a section title common in computer science. There are three sections in the Process division, comprising more than half of the article. The first, Preliminaries, explains the mathematical terms and procedures used in the current research, followed by two sections describing the current research. Testing methods and results are presented in section 6, followed by a separate Discussion, rare in engineering articles.

Finally, another electrical engineering paper is shown in Table 2.6 (Tsai et al., 2011). This paper differs from the previous two because the contribution of the work is a new mathematical procedure and its proof. Since mathematical proof does not require

Table 2.4 Section titles and length in an IPTC article, "Eight-element receiver"

Div.	*Section/Subsection titles*	*Paragraphs*	*Word count*	*%*
I	I. INTRODUCTION	3	391	
	II. EIGHT-ELEMENT PROGRAMMABLE PPA RECEIVER ARCHITECTURE	3	351	17
P	III. TECHNOLOGY, DESIGN, AND BUILDING BLOCK MEASUREMENTS			
	A. Technology	1	106	
	B. Reconfigurable Input Switching Network	9	402	
	C. Wideband Phased Array Channel	5	441	48
	D. Reconfigurable Output Combining Network	4	404	
	E. Differential-to-Single-Ended Converter	1	140	
	F. Digital Beamforming Path	1	301	
	G. Digital and Reference Current Control	2	188	
	H. Layout	1	77	
T	IV. SYSTEM-LEVEL SIMULATIONS	6	683	
	V. MEASUREMENTS	1	82	33
	A. S-Parameters, Linearity, and Noise	4	450	
	B. Isolation and Forward Coupling	3	208	
C	VI. CONCLUSION	1	109	3
Total		45	4333	100

Table 2.5 Section titles and length in an IPTC article, "Spatial query integrity"

Div.	*Section/Subsection titles*	*Para-graphs*	*Word count*	*%*
I	1 Introduction	8	1164	19
	2 Related Work	4	742	
P	3 Preliminaries	8	551	
	3.1 Voronoi Diagrams			
	3.2 Signature Aggregation			
	4 Data Transformation	4	609	
	5 Authenticating Spatial Inquiries	27	4304	
	5.1 Query Processing at the SP			53
	5.2 Signature Verification			
	5.3 Geometric Verification			
	5.3.1 kNN and Range Query Verification			
	5.3.2 RkNN Query Verification			
	5.3.3 KaNN Query Verification			
	5.3.4 Spatial Skyline Query Verification			
T	6 Experiments	13	2219	
	6.1 Experimental Settings			
	6.2 VN-Auth and MR- and MR*-trees			26
	6.3 Advanced Spatial Inquiries			
	7 Discussion	2	471	
C	8 Conclusions	1	172	2
Total		67	10248	100

Table 2.6 Section titles and length in an IPC article, "On 4-ordered 3-regular graphs"

Div.	*Section/Subsection titles*	*Para-graphs*	*Word count*	*%*
I	1. Introduction	6	595	13
P	2. 4-Ordered cells	17	1966	84
	3. Generalized honeycomb torus	24	1819	
C	4. Main result and concluding remarks	3	125	3
Total		50	4505	100

empirical testing, there is no testing division, resulting in an IPC structure. Since there is no Testing, and only a short Introduction and Conclusion, the Process division accounts for most of the article.

Several basic differences between IMRaDC and IPTC formats are illustrated in Figure 2.1.[4] Remember first that although the section names of the former are usually general, section names for the middle two divisions of IPTC tend to be specific to the

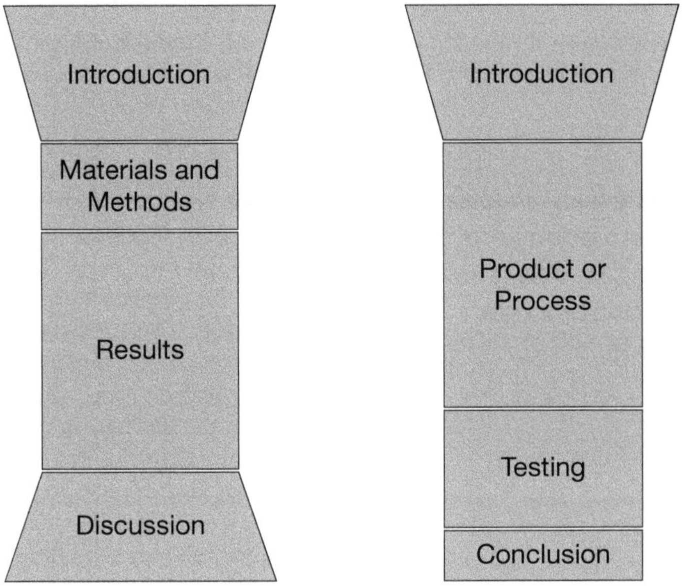

Figure 2.1 Overall shape and relative length of divisions in IMRD and IPTC formats.

research. Thus, whereas *IMRaDC is the final structure of the paper, IPTC is a scaffold to help you understand or build the structure.*

Next, the length and shape in the figure indicate the relative length and breadth of each division. As we saw above, whereas IMRD focuses on Results and Discussion, in IPTC the new Product or Process frequently accounts for half the length of the article. Furthermore, IMRD papers often have an "hourglass" configuration.[5] Starting with a broad overview of the history of the field, they gradually become more specific until they reach the focus of the current research. Then the bulk of the paper focuses on the current research, but the Conclusion frequently broadens again to applications of or future directions for the research. IPTC papers, although they also focus from broad to narrow at the beginning, tend to talk less about future extensions of the work. Again, these are general statements that may not be true for each article. For example, if an IPTC article combines two established methods, the Process division may be shorter, since both have been described before, and the Testing division longer.

2.2.3 Why a separate format?

Since IPTC could be conceived as simply another modification of the well-known IMRD structure, with more sections, more specific names, and an emphasis on the second division, the obvious question is why a new designation is needed. Taxonomy is always an ontological issue, generating a debate between "lumpers" and "splitters," whether in biological species (Rau, 2012: 125, 132) or genres of academic writing (Swales, 2004: 63–65).

Unfortunately, every recent book on academic writing has presented the IMRD format from the perspective of quantitative empirical science, in which the Results section dominates and directs the structure of the rest of the paper, with Methods relegated to a minor status. *Nothing that has been written on IMRD in the last 50 years matches the actual features of a prototypical engineering article*, to be described in detail in the next several chapters, in which the Process division is the longest and most important part of the paper. Rather than trying to broaden the deeply entrenched understanding of IMRD and then trying to distinguish IMRD$_1$ from IMRD$_2$, it is easier and clearer to propose a new name for the analogous structure, and the difference is easily understood by students (Rau, In preparation).

> Prototypical engineering articles do not match past descriptions of IMRD. Rather than trying to differentiate IMRD$_2$, I propose a new designation, IPTC.

> Science asks questions and develops explanations based on data; engineering defines problems and designs superior solutions.

The distinction between the two formats can easily be explained by considering how the goals of science and engineering differ. Science asks questions (Introduction) and seeks to develop explanations (Discussion) based on an analysis of the data (Results), whereas engineering defines problems (Introduction) and seeks to develop solutions (Product or Process) which are compared with the current best solution (Testing), definitions based on the *Next Generation Science Standards* (Rau & Antink-Meyer, 2020).

A summary of the focus of each division of the paper will serve to highlight both the similarities and an important difference between IMRD and IPTC (Figure 2.2). Note that the focus of the IMRD format is on the *data* generated and their interpretation. Therefore, the longest and most important part of the paper is usually the Results section, where a summary of those data is presented, or sometimes the Discussion section, where they are explained. On the contrary, the focus in the IPTC format is on the new *design*, the new Process or Product that is described, so this is the longest and most detailed part.

2.3 Variation in research article formats

So far we have only considered the prototypical formats of science and engineering, but even a quick look at research articles will reveal that there is substantial variation in format within a field, often within a sub-field, and even sometimes within a single journal. In this

IMRD:

Introduction: Questions to be addressed by the *data*
Materials and Methods: How the *data* were generated
Results: Summary of the *data*
Discussion: Explanation of the *data*

IPTC:

Introduction: Problem to be solved by the *design*
Product or Process: Solution proposed by the *design*
Testing: Comparison with previous *design*
Conclusion: Advantage of *design*

Figure 2.2 Summary of the focus of each division in IMRD and IPTC formats.

section I will present a framework for understanding a likely cause of that variation. Since this is more theoretical, if your exemplar articles fit the prototypical pattern you may skip this section and come back to it later, after you understand the basic structure better.

The foundation of this framework is the idea that *our traditional delineation of fields is based on topics of study, whereas the reporting format used for writing research articles is more closely aligned with research methodology*. For example, electrical engineering (EE) investigates various practical applications of electricity and magnetism. Prototypical EE studies design a physical product, but some develop a new algorithm, and thus have a reporting structure almost identical to applied mathematics, while others spend more time testing various alternatives and are similar to a typical science paper.

Fields are delineated based on topics studied, whereas the choice of reporting format is based on what works best for the research methodology.

One important corollary of this is that although IMRD is the prototypical paradigm for science and IPTC the prototypical paradigm for engineering, not all research in a field will follow the prototype. For example, research in biomechanical and chemical engineering and materials science frequently follows the IMRD reporting structure. On the other hand, much of the physics literature utilizes IPTC.

Science has not always used the IMRD format, and it is by no means universal in natural science research articles, let alone engineering or other disciplines. Even a quick look at journals from different fields reveals at least three basic types of research, each focusing on a different division of the paradigmatic IMRD framework, as summarized in Table 2.7. Each type appears to be associated with a different reporting format, all of which have previously been classified as variants of IMRD.

Table 2.7 Focus of different types of research

Research method	*Research focus*	*IMRD focus* *Reporting format*
Methodological	Solution	Methods
Non-empirical	Theoretical, mathematical, logical	IPC
Design	Product, process	IPTC
Empirical	Data	Results
Quantitative	Repeatable, statistical	IMRD (or IMRaDC)
Qualitative	Non-repeatable, situation-specific	IMRaDC (or IMRD)
Analytical	Interpretation	Discussion
Inductive	Integration, generalization	IMRaDC, Recursive
Deductive	Evaluation, comparison	IRDC

Most previous work on genre analysis has considered only differences between various fields of research, rather than research methodology. The framework proposed here may help explain what were considered anomalies, such as the great difference occasionally seen between articles within the same field or science fields such as seismology in which many articles follow a recursive structure. Work to confirm and refine the framework, especially for analytical research (more common in the humanities and social sciences), is ongoing (Rau, In preparation).

2.3.1 Formats in the natural sciences

Prototypical research in the natural sciences at present uses a *quantitative* methodology. IMRD is the most common reporting format in the natural sciences because it is uniquely suited to this type of research. In quantitative studies, new data are generated, often to test a prediction and thus its underlying hypothesis (see section 8.5.2). After an Introduction, the Method of collecting the data is described, then the data are reported in the Results section and explained in the Discussion. This is the simplest and most direct method for reporting a single quantitative empirical study, and thus has become the standard reporting format. Some fields use a modified structure, with a combined Results and Discussion and separate Conclusion.

However, many branches of science employ other research methods. Theoretical research, common in many branches of physics, may be entirely *non-empirical*, in which case there may be no Results or Discussion, only an Introduction followed by a mathematical proof or logical argument and a Conclusion. When a new method is presented for gathering or analyzing data in empirical fields, the focus of the article will obviously be on the

Although the quantitative paradigm establishes IMRD as the prototypical format for natural science, other reporting formats are suitable for different research methods.

Methods, with a short section comparing the new method with existing methods, leading to a structure essentially identical to IPTC.

On the other end of the scale, fields such as epidemiology frequently use a *qualitative* paradigm of observational research to study objects or events that cannot be controlled or repeated. Some research uses an *inductive* approach, examining and trying to consolidate the results of earlier primary research studies. This is common in environmental science, seismology, and other fields exploring large-scale phenomena, where it is often impossible to collect sufficient data in a single study. Since the focus is on analysis and interpretation, the traditional role of the Discussion, these fields often use the recursive reporting structure typical of social science. Some disciplines that make extensive use of *deductive* or descriptive research, such as traditional botany and taxonomy where a new type is compared with known types, have a very different reporting structure, totally lacking a Methods section.

2.3.2 Formats in engineering and mathematics

In engineering the goal is to find a *design* solution to a practical problem, so in the prototypical case the focus is on how the new Product or Process was developed and what makes it distinct. This leads to the typical IPTC format.

However, some sub-fields of electrical and mechanical engineering may include whole sections in which mathematical proof is used, and are thus largely *non-empirical*, with little or no Testing division. On the other hand, engineering studies using quantitative techniques, common in chemical engineering, will follow an IMRaDC format when reporting. Some fields, such as civil engineering, may use either format depending on whether the research is more oriented toward basic materials or their practical use.

> Since research methods vary, even within a field, some engineering articles are written in IPTC, others in IMRD or mathematics format.

Mathematics articles, apart from the Introduction, typically follow the conventions of a *non-empirical*, logical deductive argument to present a new way of solving a problem or an extension of a previous solution. As with theoretical science, since the emphasis is on logical rather than empirical verification, there may be no Results and Discussion.

2.3.3 Formats in the social sciences

The social sciences are equally varied. Given the difficulty of collecting large data samples, some work in social sciences is secondary research, so the research type is *analytical*. The structure of these articles is similar to that found in the humanities, with extensive quoting and comparison of different primary sources.[6] The focus of the article is on analysis, either comparison of the present study with previous studies, compilation of the results of other studies, or other topics commonly found in the Discussion. In many cases the order within an individual section seems to follow IMRD format. The authors will discuss literature related to a particular method, the results that have been obtained, and their interpretation

Likewise, social science research uses various reporting formats, depending on the research methodology.

of those data. The same process may be repeated recursively (cyclically) in subsequent sections for other types or sources of data, followed by an overall final analysis.

Nevertheless, not all social science literature is analytical. Primary research from a *qualitative* paradigm places far more emphasis on describing the process of data collection and interpretation. Primary research in the social sciences done from a *quantitative* perspective tends to be written up in IMRD format, just as in the natural sciences. Theoretical studies in economics and related fields may employ extensive mathematics and use a *non-empirical* format like theoretical science, without a Results section. Some branches of linguistics, such as formal semantics, employ a similar structure, given their emphasis on logical consistency rather than empirical verification. Given the variety of reporting formats, the technique described in Chapters 3–9 for analyzing exemplar articles in the science and engineering literature is equally useful for the social sciences.[7]

Although science articles typically use IMRD format, while engineering uses IPTC, there are exceptions. Exemplars will reveal the format(s) used by each journal.

2.4 Determining exemplar article structure

In section 2.2 I introduced the prototypical formats for science and engineering articles, but individual fields or articles may differ, for reasons explained in section 2.3. In this section I will give more details on the difference between the formats so you can decide which is used by your exemplar articles and designate the division breaks.

First, based on the number of sections and the section titles, decide which format your article seems to follow, then assign sections to divisions. This is easy with IMRD, since general section titles are used by many articles. Almost all IPTC articles have sections entitled Introduction and Conclusion, so assigning those to the respective divisions is easy, but other sections require more explanation.

A section with a title like Related Work, Previous Work, Background, or Motivation is usually part of the Introduction division.

If there is a section immediately after the Introduction section with a title such as Related Work, Previous Work, Background, Motivation, or something similar, this is usually part of the Introduction division. As we will see in the next chapter, its main purpose is to further establish the need for the present work. Occasionally this will be placed at the end of the article, as the Method division may be in IMRD, but logically it is still part of the Introduction division.

> In IPTC, identify the Testing division based on words that indicate testing, results, experiment, measurement, simulation, or comparison.

The Testing division addresses the claim that the proposed solution is better than other existing solutions, which includes both testing the solution and comparing it with others. Section titles in this division frequently include the words testing, results, experiment, measurement, simulation, comparison, or other words that relatively unambiguously identify the division. Sometimes there will be more than one section, either two types of tests, a description of tests and results, or a separate Discussion section. Articles where the main contribution is a new mathematical proof or algorithm may have no Testing division.

> The Process division, between Introduction and Testing, includes sections with titles like Preliminaries, Problem Definition, or names specific to the current research.

What remains between the Introduction and the Testing division is the Process division.[8] This may include sections with titles like Preliminaries or Problem Definition that give background or vocabulary necessary to understand the present work, in addition to titles specific to the current research. If a section includes many sentences with "we" or "this paper," it describes the current research rather than previous research, the focus of the Introduction. Sometimes the Process division will contain intermediate testing necessary to the development of the new design. In this case there may be little or no Testing at the end, so the Process and Testing divisions may be combined. This will become clearer in subsequent chapters as we look in detail at each division, so you can change your designations later if necessary.

Exercise 2.1 Overall structure

(1) Identify the format (IMRD, IMRaDC, IPTC, or IPC) and break points between the divisions for your first exemplar article.
(2) Create a table showing the section titles, paragraph numbers, and word count for each division in your exemplar article, following a format similar to Tables 2.1–2.6.

Hint: List the section numbers and titles in the format used by the journal, to get used to it and note differences between journals. If your journal does not use numbers, you will have to pay close attention to font size and location to determine the level of the headings. If the PDF has bookmarks, the levels can easily be seen there.

Hint: Word count will be useful not only to determine the relative length of each division, but also the total length, giving you a goal for your own writing. Export the text from PDF into Word, or copy each section into a Word document, then use the Word Count tool.

Hint: It may be hard to get an accurate word count for articles with large numbers of non-text features like equations. Give the best estimate you can.

Hint: Should you count the words separately for each section and subsection? I did for most of the tables in section 2.3, but not for Table 2.4 because the large number of sub-subsections was not a common feature in that journal. Your goal is to establish the typical pattern.

(3) Note any difficulties, for example in getting an accurate word count, just as you would any other problem with data collection.

Hint: Make sure you keep your notes from all these Exercises, as they will be helpful when you do your own writing and rewriting.

Evaluate your understanding

(1) What is the difference between a "section" and a "division" of a research article?
(2) How do section titles differ in IMRD and IPTC?
(3) Why does the relative length of divisions differ in IMRD and IPTC?
(4) Why are there differences in reporting format within a field?

Homework

(1) Repeat Exercise 2.1 for your other two exemplar articles. Number and give titles to your tables.

Hint: Think about how to best organize your information to make it easy to understand. The way you collect and record the data may not be the best way to present it. This is true for all research. For example, instead of using a table to show the total word count for each division, you could use a graph or figure to indicate the relative proportions. What would be the advantage of each? Which would convey the important information most clearly?

Hint: If you have more than one table or figure, how should they be numbered? Should they be, for example, Table 1, Table 2, and Table 3, or Table 1a, 1b, and 1c, or Table 1.1, 1.2, and 1.3 (that is the format used in this book, but why)? What would make it easier for the reader to see that they are connected? Which is more similar to your exemplar articles?

Hint: The table title should describe the contents of the table, not just list the name of the article. For example, "Table 1a: Overall structure of exemplar article 1."

Hint: Eliminate extra white space from your tables to make them easier to read.

Teaching tip: It may be useful to look at students' work on one article first, before asking them to complete the same exercises for the other two articles. Your feedback on the first will reduce both their work and yours by minimizing errors, and gives a second chance to see if they have understood the chapter.

Teaching tip: Have students compare their work, either in class as they do the exercises or after completion of the homework, so they can see how much variation there is from one journal or field to another. Also have them compare table formats, table titles, and the overall organization of their summaries, to determine what works best and why.

Teaching tip: As students learn about their exemplar articles they will be asked to write a description step by step. This will give them practice with writing in stages, rewriting, making tables, and various skills that will be taught deliberately later in the book.

Teaching tip: Ideas from section 24.3 could be discussed at this point, to teach students how to format and title a table effectively.

Notes

1 The sample articles were selected not for being prototypical, but by historical contingency. They were submitted by my students as exemplar articles during the years I was developing the material in this book.
2 IMRD and IMRaDC are often jointly called IMRaD.
3 In the journal *Carbon*, the Methods division is commonly given the section title Experimental, Experimental Procedure, Experimental Details, or Experimental Setup and Methodology.
4 Earlier versions of Figures 2.1 and 2.2 appear in Rau and Antink-Meyer (2020).
5 The hourglass shape of the IMRD format has been noted by many authors, including Cargill and O'Connor (2013) and Glasman-Deal (2010), but none I have seen distinguish the relative length of the divisions.
6 This has important implications for academic writing classes. While students in the humanities and social sciences can write a research paper summarizing other work in a format very similar to that of a research article in the field, this is not the case in science and engineering, where such a paper would be similar to a thesis literature review, but not to any part of an academic article.
7 I have had several social science students in my classes who have used the method presented in this textbook for their work, although they needed additional help to identify the recursive nature of writing in their discipline.
8 Occasionally presentation of each portion of a design is followed by testing of that portion in the same section, leading to a combined Process and Testing division, but this seems to be rare.

References

BAZERMAN, C. 1983. Reporting the experiment: The changing account of scientific doings in the *Philosophical Transactions of the Royal Society*, 1665–1800 (mimeo).

CARGILL, M. & O'CONNOR, P. 2013. *Writing scientific research articles: Strategy and steps*, New York, John Wiley & Sons.

GLASMAN-DEAL, H. 2010. *Science research writing for non-native speakers of English*, Singapore, World Scientific.

HU, L., KU, W.-S., BAKIRAS, S. & SHAHABI, C. 2013. Spatial query integrity with Voronoi neighbors. *IEEE Transactions on Knowledge and Data Engineering*, 25, 863–876.

KLEINHEINZ, G. T., MCDERMOTT, C. M., HUGHES, S. & BROWN, A. 2009. Effects of rainfall on *E. coli* concentrations at Door County, Wisconsin beaches. *International Journal of Microbiology*, https://doi.org/10.1155/2009/876050.

PENG, X., ZHAO, S., OMASTA, T. J., ROLLER, J. M. & MUSTAIN, W. E. 2017. Activity and durability of Pt-Ni nanocage electocatalysts (sic) in proton exchange membrane fuel cells. *Applied Catalysis B: Environmental*, 203, 927–935.

RAU, G. 2012. *Mapping the origins debate: Six models of the beginning of everything*, Downers Grove, IL, InterVarsity Press.
RAU, G. In preparation. IPTC: A prototypical format for engineering research articles.
RAU, G. & ANTINK-MEYER, A. 2020. Distinguishing between science, engineering and technology. *In*: MCCOMAS, W. & ORAMOUS, J. (eds.) *Nature of science in science instruction: Rationales and strategies*, New York, Springer.
SAYGINER, M. & REBEIZ, G. M. 2016. An eight-element 2–16-GHz programmable phased array receiver with one, two, or four simultaneous beams in SiGe BiCMOS. *IEEE Transactions on Microwave Theory Techniques*, 64, 4585–4597.
SWALES, J. M. 1990. *Genre analysis: English in academic and research settings*, Cambridge, Cambridge University Press.
SWALES, J. M. 2004. *Research genres: Exploration and analysis*, Cambridge, Cambridge University Press.
TSAI, M., TSAI, T.-H., TAN, J. J. & HSU, L.-H. 2011. On 4-ordered 3-regular graphs. *Mathematical and Computer Modelling*, 54, 1613–1619.

Chapter 3

Argument structure of research articles

3.1 One long argument

When Charles Darwin wrote his classic, *On the Origin of Species*, he famously described it as "one long argument." In the first sentence of the final chapter (XV) he states, "As this whole volume is one long argument, it may be convenient to the reader to have the leading facts and inferences briefly recapitulated" (Darwin, 1859).[1] Actually it is a complex series of arguments, nested one inside another, with claims, evidence, and reasoning interspersed. Each chapter presents one general claim. Paragraphs generally begin with a topic sentence that lays out the specific claim supported in the remainder of that paragraph, unless the purpose of that paragraph is to provide general background before stating the next claim. Part of the reason his work had such a great impact was his command of argument structure, which is essential to any scientific or engineering writing.

Broadly speaking, any research based on empirical evidence can be considered to be scientific. In this sense both social science and engineering, along with natural science, can be considered to be scientific research. All scientific study involves four main elements, although the relative importance of each varies:

(1) Presuppositions;
(2) Evidence;
(3) Inference;
(4) Argumentation.

> Even though it may appear to be an objective report of the data, scientific writing is in fact an argument supporting a particular conclusion.

The most important point for our purposes in this book is to understand that even though it may appear to be an objective report of the data, *scientific or engineering writing is not so much a report as it is an argument.* With regard to research articles, in a typical natural science article you are trying to demonstrate the claim that you have collected reliable data (evidence), interpreted it in a reasonable way (inferences), and thus have a valid explanation for a phenomenon. In an engineering article you are trying to show that you have a better solution to a problem than any that has been presented to date. Other genres support different claims, as we will see in Part 3 of this book.

Almost all scientific study is based on some form of *empirical evidence.*[2] Empirical evidence is something that can be observed by our senses, either directly or indirectly.

In primary research the researcher collects the evidence; in secondary research the researcher relies on the observations of others, either contemporary or historical records. Although philosophers of science have shown that collection of evidence and what is accepted as evidence can be affected by the expectations of the researcher, evidence is generally considered to be fact.

Science also involves *logical inference*. Inferences are interpretations of the evidence based on reasoning. Predictions of what we ought to observe in specific situations can be made based on general principles (deductive inference) and then tested to see if those predictions are correct, or general conclusions made based on specific evidence (inductive inference). Deductive inference and prediction testing are far more common in normal science activities, while inductive inference, coming up with a new theory to explain a vast array of diverse data, can lead to major paradigm shifts. In the absence of a clear hypothesis, abductive inference chooses the most likely of several possible explanations.

The importance of *presuppositions* in science is often overlooked, but here I will only say that science cannot be done without certain philosophical presuppositions about the reality and orderliness of the world and our ability to understand it. Other presuppositions are specific to one theoretical model. These presuppositions may affect the way we perceive and interpret the data more than we like to admit, and incorrect presuppositions lead to faulty conclusions (Gauch, 2003; Rau, 2012).[3]

Writing in science and engineering almost never mentions presuppositions, which can cause problems in areas where the topic impinges on values, ethics, or other areas science cannot address.[4] The degree to which the writing is based on evidence or inference depends on the field and the question studied.

No matter what, *all academic and technical writing is (or should be) structured as a series of arguments*. Argument in this context does not mean disagreement, but a logical argument, where a claim is made and supported by evidence and reasoning. In the US this is often taught as "claim, evidence, reasoning," which is based on the classic formulation "grounds, claim, warrant" (Toulmin, 1958), with the order of the first two terms reversed.

When analyzing your exemplar articles, it is important to remember that a research article is not just a report, but an argument, even though in science and engineering it may be harder to detect than in the analytical reporting style of the social sciences. Thus it is important to look not only at the format of the article, but also the structure of the argument. It may be hard to see the argument structure at first because there is frequently a nested structure of claims within claims, and each claim may be stated before or after the evidence supporting that claim. Nevertheless, recognizing what types of claims and support to look for is an important first step.

3.2 Three implicit claims

> These implicit claims direct the overall argument but are rarely stated directly.

Within any article, there will be several separate but related arguments, each making a claim that requires support.[5] In my experience, science and engineering articles make three general implicit claims, as shown in Figure 3.1.[6] The claims for the two

Implicit claims in science and engineering research articles

Science (IMRD)

1. You have identified an important unanswered question
2. You have gathered data in a valid, reliable way
3. You have a good explanation for your data

Engineering (IPTC)

1. You have identified an important unresolved problem
2. You have designed a workable solution to the problem
3. Your solution is better than existing solutions

Figure 3.1 Implicit claims in science and engineering research articles.

formats are similar, differing only because science asks questions and seeks to develop explanations based on an analysis of the data, while engineering defines problems and seeks to develop solutions that are better than the current best solution, as described in section 2.3.3. These three implicit claims can be broken into ten component claims, described in the next subsection. In the next three chapters you will look at exemplar articles in your field to try to identify each of these claims.

The first claim, supported in the Introduction, has two parts—that the work is important and that it has not been done. If it is unimportant or an adequate explanation or solution is already available, the research is not needed, so the Introduction establishes why an article should be published.

The second claim is implied in the way the Methods or Process divisions are written. Statements in the Methods about following established procedures, care in measurement, and replication establish the implicit claim that the data were collected in a valid, reliable way. Detailed description of the new product or process similarly supports the implicit claim that you have designed a workable solution.

In IMRD, supporting evidence for the third claim, that you have a good explanation of your data, is placed in the Results, but the logical argument occurs in the Discussion.[7] In IPTC, the claim that your design is superior to existing designs is supported in the Testing division.

You may have noticed that the arguments are closely related to the structure of the article, as summarized in Table 3.1. In both formats, each claim is closely associated with one division, although in IMRD evidence and reasoning for the third claim are split between the Results and Discussion, respectively. *Because the Results and Discussion together support a single claim, we will treat the two of them together throughout this book.* Even though these claims are rarely stated explicitly, the fixity of the overall order of a research article makes it clear that both scientists and engineers feel it necessary to address these issues.

Table 3.1 Locations of support (evidence and reasoning) for implicit claims

Implicit claim	*Support location*	
	IMRD	*IPTC*
1	I	I
2	M	P
3	R & D	T

3.3 Ten common component claims

Within these three implicit claims are ten common component claims, which I will henceforth refer to as components, present in most research articles (Figure 3.2). We will introduce the components here, then expand on each in Chapters 4–6.

The first common component, *Importance*, is obviously related to the first part of the first implicit claim, establishing the importance of the research, but the degree of support necessary varies. For a new area of research, more reasoning will be necessary. For established questions or problems in the field, citing the authors who have already worked on it is sufficient. As a result, this tends to be less prominent in IPTC, where each paper is likely to present one small improvement on a problem many research groups are already addressing.

The second common component, *Need*, addresses the second part of the first implicit claim in each format, that the question is unanswered or the problem unresolved. This is often expressed as a limitation of previous studies, following a description of what has already been done. Detailed support for the first and second components may be placed in a second section entitled Related Work, Background, or Motivation.

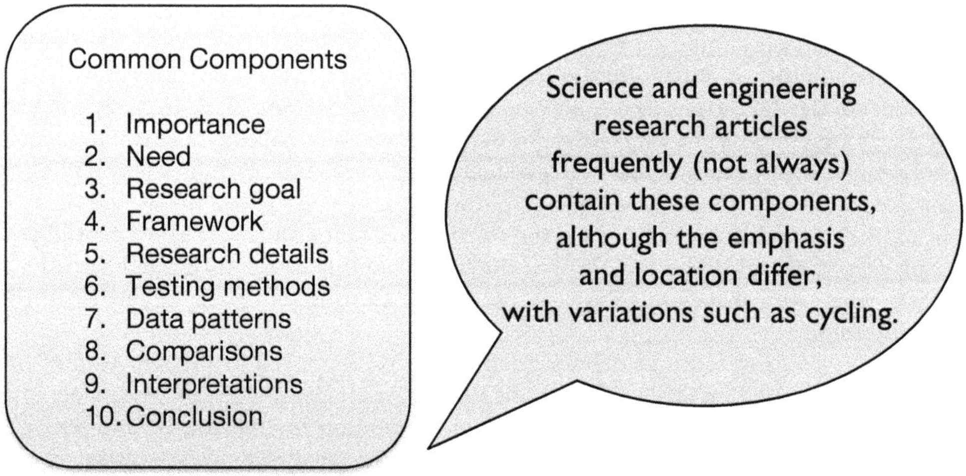

Figure 3.2 Ten common components of science and engineering writing.

It may seem odd to consider the third component, *Research goal*, as a claim, when it is often stated as either questions or aims. Nevertheless, it is an essential component of any research article, and makes an implied claim that the questions are answered in the paper or the aims achieved.

Framework involves justification of the research methodology chosen. Even if not stated explicitly, the implication is that the researchers have considered different ways of addressing the question or solving the problem and believe the model, theory, or framework used to be the best for the purpose.

The fifth component, *Research details*, likewise should not merely be a list of what was done but should be written in a way that addresses the second implicit claim, that the data are valid and reliable or the solution workable. The care shown by writing a detailed description is taken as evidence of similar attention to detail during the research. The same is true of the sixth component, *Testing methods*.

Raw data are of little use in research until a *Data pattern* can be described, or exceptions to a general pattern noticed. Thus in the Results division the focus is not on reporting the data itself, but patterns in the data, which are often easier to see graphically, explaining the abundance of figures and tables in this division. The claim that there is a pattern is often confirmed by statistics, triangulation with other data sources, or another method.

Many different types of *Comparisons* are used in science and engineering research, including comparisons with predictions, theoretical ideals, or previous results. These are more obviously claims that, along with *Interpretation* of the data, must be supported by both evidence and reasoning.

The final component, *Conclusion*, is usually stated in the last one or two paragraphs, whether that is the end of the Discussion or a separate Conclusion, either as answers to the research questions or a statement that the aims have been accomplished. Support for this claim is spread throughout the rest of the document.

A summary of the components and differences between IMRD and IPTC where important is shown in Table 3.2.[8] More detailed descriptions of each will be provided in subsequent chapters and are summarized in the generalized component list in Appendix 1.

> **The order of components is similar in IMRD and IPTC, but the division structure differs.**

Although all ten components are common in both IMRD and IPTC, and tend to follow the same order, they are not necessarily found in the same divisions, as shown in Table 3.3. This is largely due to the differences between the three implicit claims in the two formats.

One difference between IMRD and IPTC is the location of the research framework, the model or theoretical framework on which the current research is built. Since there are so many unanswered questions in science, there is a large available research space. It is easier to show the importance and need for your research if you can tie your work to an established research tradition, thus the research framework is often part of the Introduction, frequently implied based on the citation and evaluation of previous research. In engineering the research space is much more crowded, with many people trying to find a better solution to the same problem. Thus it is more important in the

Table 3.2 Ten common components of science and engineering research articles

Component	*Sub-components common in science and engineering*
1 Importance	This research is important to society This topic is important to researchers in the field
2 Need	A gap exists in current knowledge or understanding (IMRD) The current best solution is limited or less than ideal (IPTC)
3 Research goal	We answer a question or improve understanding (IMRD) We propose a better solution to a problem (IPTC)
4 Framework	The research is based on an accepted model or framework This is the best model or method to follow for our research
5 Research details	Care was taken to ensure good results (IMRD) A workable solution was developed (IPTC)
6 Testing methods	We can predict results based on the model Testing followed verifiable procedures
7 Data patterns	A pattern can be discerned in the data Exceptions to the pattern can be identified
8 Comparisons	Data [support/question] previous work Data [conform to/differ from] expectations
9 Interpretations	Data are best interpreted in a certain way There is a reasonable cause for the data
10 Conclusion	The question has been answered or the aim achieved The solution is an improvement on the current best design

Table 3.3 Typical location of components in science and engineering research articles

IMRD (Science)		*IPTC (Engineering)*	
Section	*Components*	*Division*	*Components*
Introduction	Importance Need for work Research goal	**I**ntroduction	Importance Need for work Research goal
	(Framework)*	**P**roduct or Process	Framework Research details
Materials and Methods	Research details Testing methods	**T**esting	Testing methods Data patterns Comparisons Interpretations
Results	Data patterns		
Discussion	Comparisons Interpretations Conclusion	**C**onclusion	Conclusion

*Often implied by the author's attitude toward previous work cited; see section 5.3

Introduction to show the limitations of previous work—the problem that remains—whereas the framework for the present work is frequently discussed at the beginning of the Process division as background for the new modification.

The other main difference is in the location of the Testing method. In IMRD the Testing method is usually more closely related to the validity of the data (claim 2) than the interpretation of the data (claim 3). On the other hand, in IPTC the Testing method is associated with demonstrating superiority over existing solutions (claim 3), more than proving the solution is workable (claim 2).

Exercise 3.1 Locating components

(1) Try to identify the main component or components in each section of your exemplar articles.

Hint: Use Table 3.3 as a guide, but remember that your article may differ.

> *Teaching tip: Argument structure of each division will be discussed further in Chapters 4–6. At this point students should get the idea that articles are written as a series of arguments and which components tend to appear in each division.*

3.4 Supporting the claims

> Each claim must be supported by evidence, either from past or present research, and reasoning.

Supporting a claim requires evidence. This has not changed since the English philosopher William Kingdon Clifford said, "To sum up: it is wrong always, everywhere, and for anyone, to believe anything upon insufficient evidence" (Clifford, 1877). Thus it is essential to support every claim with sufficient evidence, either from past or present research. Support also requires reasoning. Reasoning involves inference and evaluation, but also employs various rhetorical devices. The ideas introduced here will be expanded in Chapters 7–9.

As shown in Table 3.4, there are two basic sources of evidence in scientific writing: what is already known and new evidence from the present research. Anything well known in a field, affirmed by many previous authors, can be used as evidence without citation. Specific research findings presented by a single author can be used as evidence by citing the original publication. Publication implies that the results have already been accepted by at least some portion of an academic community.

Past research is generally used to support claims at the beginning of the paper: Importance, Need, and Framework. Research details and Testing methods usually build on previous research as well. Commonly used methods do not require a citation, but methods used by only a few previous researchers do.

Table 3.4 Two sources of evidence in science and engineering articles

Evidence	*Examples*
1 Accepted (past)	General knowledge, definitions (no citation) Theoretical basis, specific research findings (citation)
2 New (present)	Detailed description of current research procedures Data (text, graphics, examples, statistics)

Details of the present design and testing procedures, often including illustrations in IPTC, are necessary to support the reliability of the data in the present work and thus also constitute evidence. Data patterns show evidence collected during the present research and used thereafter to support the claims. Past research may be brought in again when doing the Comparisons or Interpretation.

There are two basic types of reasoning used in scientific writing: logical and evaluative, as shown in Table 3.5. Reasoning also involves ensuring the organization is clear to the reader.

As explained in section 3.1, all scientific work involves at least one type of logical inference, with deductive being the most common. Science also involves interpretation of evidence, new and old. Organizational strategies can help a reader follow the author's reasoning. Papers in which the reasoning is clear are much easier to read than those in which it is not. Thus the instructions for rewriting in Chapters 21–23 focus on how to improve the argument structure and make the reasoning clearer to your audience.

In the next Part you will look at both the claims and support in more detail. In Chapter 4 you will learn how to identify components in the Introduction division. Chapter 5 will explore the second main division, either Methods (IMRD) or Process (IPTC), which are very different in the two formats, followed by a chapter on the last two divisions, which report the data summary and analysis. Chapters 7 and 8 show how evidence from past and present research, respectively, is used to support each component claim, before concluding Part 2 with a discussion of reasoning and sequencing strategies in Chapter 9.

Exercise 3.2 Writing and supporting a claim

(1) Practice writing and supporting a simple claim. Based on your homework for Chapter 2, claim which format your first exemplar follows, and the support for that. If there are any unusual features, note them as well.

Hint: For example (based on the article in Table 2.5), you might write, "This article is written in IPTC format. As shown in Table 1a, there are eight sections, two of which (4 and 5) have titles specific to the research. More than half of the article is in the Process division. It has sections named Related Work and Preliminaries, which are found in almost every paper in this journal. The article is unusual because it has a separate Discussion section in the Testing division."

Teaching tip: This is support from present evidence.

Table 3.5 Types of reasoning in science and engineering articles

Reasoning	*Examples*
1 Logical	Deductive (general theory tested by specific observations) Inductive (specific observations to general conclusion) Abductive (most likely of several possible explanations)
2 Evaluative	Evaluation of previous research findings Interpretation and explanation of current data
3 Organization	Introductory or concluding summaries Referring to prior or following arguments

Evaluate your understanding

(1) Why is it important to understand the argument structure of research articles?
(2) What implicit claims are commonly made in science and engineering articles?
(3) What components are found in most science and engineering writing?
(4) Why are some components found in different divisions in IMRD and IPTC?
(5) How are claims supported in research articles?

Homework

(1) Repeat Exercise 3.2, combining information from all three exemplar articles into a single summary of the general structure in your field. Do they follow IMRD or IPTC format? How many divisions are there, and how are they named? Which division is the longest? Refer to each table or figure as support. Note any unusual features or difficulties.

Hint: Would it be better to write this summary in one paragraph or one for each article? Which should be placed first, the written summary or the tables, or should the two be interspersed? What would make it easier to read?

Teaching tip: Emphasize that students should combine information from all three articles into one general description. Peer review helps students consider different formats for the summary.

Notes

1 This sentence mentions all the elements of science except presuppositions: facts (evidence), inference, and argument.
2 The exception being theoretical work, which may be based on mathematical proof and precede empirical confirmation, as in the case of Einstein's laws of relativity.
3 This is closely related to what has been called motivated cognition or motivated reasoning.
4 In fact, this is quite common, e.g. artificial intelligence (AI), environmental issues, genetic engineering, etc.
5 This is a broader use of claim than the "knowledge claim" discussed in linguistics literature (e.g. Myers, 1992), which is similar to what is in this book called the first "implicit claim" and closely associated with the "research goal."
6 Ironically, I am making this claim, but can only back it up with reasoning and anecdotal evidence. That is why I need to soften the claim by adding, "In my experience." Whether papers in science and engineering actually make these claims has not been documented in the applied linguistics literature, but would make a wonderful topic, to confirm or disprove my anecdotal observations by a systematic study.
7 Thus neither the Results (evidence) nor the Discussion (logic) by itself constitutes a complete argument. The Discussion must frequently refer back to the Results, as well as to data or analysis from other authors. This is why it is difficult both to describe what should go in the Discussion and to write it.
8 Earlier versions of Tables 3.2–3.3 appear in Rau and Antink-Meyer (2020).

References

CLIFFORD, W. K. 1877. The ethics of belief, *Contemporary Review*, Vol. 29 (December 1876–March 1877): 289–309. *In*: VELASQUEZ, M. (ed.) *Philosophy: A text with readings* (2017), 13 ed., Boston, MA, Cengage Learning.

DARWIN, C. 1859. *The origin of species*, Mentor Edition.

GAUCH, H. G. J. 2003. *Scientific method in practice*, Cambridge, Cambridge University Press.

MYERS, G. 1992. 'In this paper we report . . .': Speech acts and scientific facts. *Journal of Pragmatics*, 17, 295–313.

RAU, G. 2012. *Mapping the origins debate: Six models of the beginning of everything*, Downers Grove, IL, InterVarsity Press.

RAU, G. & ANTINK-MEYER, A. 2020. Distinguishing between science, engineering and technology. *In*: MCCOMAS, W. & ORAMOUS, J. (eds.) *Nature of science in science instruction: Rationales and strategies*, New York, Springer.

TOULMIN, S. 1958. *The uses of argument*, Cambridge, Cambridge University Press.

Part 2

Argument structure in exemplar articles

Chapter 4

Establishing the importance of and need for the research

4.1 Joining a community

In this and the following chapters you will examine your exemplar articles to determine the unstated expectations for an article in your field. I will give guidelines about things that are *commonly* found in science and engineering research articles, and the way such articles are *typically* structured, but will also show you how to decide whether your articles *actually* follow that structure or some other.

The first step in building a house is to lay a solid foundation. Although the foundation determines the shape of the building, it may be hard to see when the structure is complete. In a new community many houses may be constructed on a similar basic plan within a short period of time. In an older community some new houses built on empty lots may imitate the older style while others choose the latest style, but sometimes the community sets standards limiting the type of construction allowed.

Analogously, all research is built on an underlying framework. It influences every aspect of the research, from the planning through the final writeup, but may be hard to see in the published article. In science it determines what hypotheses are tested, what type of evidence is sought, what methods are used to obtain it, and what types of interpretations are expected. In engineering the problem identification, design methods, and standards are largely defined by the framework chosen. Without a solid foundation the research, like a house, will be weak and unlikely to stand long.

Over time a community grows up around each research framework. In new areas of research much of the work will follow a few basic designs, but in older areas of research there may be a mix of old and new frameworks. As with houses, in research there are differences of opinion on the merit of different frameworks, and potentially animosity if the presuppositions differ.[1] Thus the framework not only guides the research, but also establishes a connection with an academic community.

> A researcher must demonstrate understanding of the history and terminology of a research community before adding to it.

In the Introduction a researcher begins to stake a claim in a community. Just like a physical community, each academic community has a common language and history, and in some cases limits on how much variation is acceptable. Thus, to join and contribute to a community, researchers must show that they can use the terminology, are familiar with the past research, and fit the expectations of that community.

Each component in the Introduction contributes to this goal. In the following sections I will show you how to identify each component present in the Introduction of your exemplar articles, but first we will look at which components are likely to be present and the reason for inclusion of each.

4.2 Claim 1: importance of and need for your work

Each paragraph in a research article should contribute to the implicit general claim being made in that division. Recall that the purpose of the Introduction is to show that:

1 You have identified an important unanswered question (IMRD)

or

1 You have identified an important unresolved problem (IPTC).

> The components Importance (1), Need (2), and Research Goal (3) are found in almost every Introduction, from which the Framework (4) may be inferred.

The current research must add something valued by the research community. This is why the first three components, Importance (1), Need (2), and Research Goal (3), are nearly ubiquitous in the Introduction to any research article, as many previous works on academic writing have noted, using different names.[2] By reading the Introduction, experienced researchers in a field will be able to discern the Framework (4), even if it is not stated explicitly, just as the foundation of a house may be largely underground but can be inferred from what is visible.

Most science and engineering research articles begin with a statement of the importance of the research, either to society or the field or both. In science several paragraphs may be devoted to this component, with extensive description of past research, while in engineering a single sentence may suffice, for reasons discussed below.

After establishing the motivation for the work, the author must show that no adequate answer or solution has been proposed, so there is a need for the present work. To do this, the author must demonstrate awareness of past research in the field. Limitations of previous research point to a gap in the research that the present research goal can fill. This component tends to be more prominent in engineering.

Every article must clearly state the research goal, which indicates the direction the authors will take to provide a new answer or solution. As we will see in section 4.6, this goal directs not only the remainder of the article, but also the preceding descriptions of importance and need. By stating that goal, you indicate that you have identified an important unsolved issue. Thus the first three components are found in almost all Introductions to academic journals, as summarized in the implicit claim.

The framework can be identified by what past research is cited and the author's attitude toward it. Sometimes it may be implied simply by where it is published, if the journal has a very limited scope. If a journal has a broader audience it may be necessary to specify the framework in either the first or second division.

4.3 Component analysis and component markers

> In English, each paragraph should have a single purpose that clearly contributes to the argument. This is often stated in a topic sentence.

Each component should be evident in the paragraph structure. Every paragraph in English should have a single purpose and discuss a single idea. It is considered good writing style in English for each paragraph to have a *topic sentence* which summarizes that main idea, usually the first sentence of the paragraph, to make it easier for readers to follow the argument. Thus in the following analysis we will attempt to determine the purpose of each paragraph.[3]

It is important to note that paragraphs in English tend to be shorter than paragraphs in Asian writing.[4] For example, one sentence in Chinese academic writing may be about the same length as an English paragraph, while a Chinese paragraph can be as long as an entire section in English. If an author does not follow the expectations for English writing, some paragraphs may include more than one component.

> Recognizing common component markers can help you both read and write research articles more quickly and effectively.

For each component I will point out *component markers* that frequently occur in science and engineering articles.[5] *Component markers are phrases or groups of words used by authors to draw attention to key components of an article.* Learning to recognize these markers can help you both read and write more effectively.

Component markers help you skim an article to look for certain information or identify the focus of a paragraph. They are often found in topic sentences, making them easier to find, but may also occur later in a paragraph.

As you write, you can add these markers to the specifics of your research.[6] This helps you write effective sentences related to each main component. Furthermore, it helps your readers understand your article better by highlighting essential information.

4.4 Component 1: importance

Most research articles begin with a statement of importance, either a broad statement of importance to society or the importance of the question to researchers, or both. These are

1) Importance

a. General importance
   Claim: This research is important to society
   Support: Often general knowledge (to those in the field)
b. Specific importance
   Claim: This topic is important to researchers in the field
   Support: Past research (citations)

Figure 4.1 Common sub-components of Component 1: importance.

> The more related research an author can cite, the easier it is to establish the importance of the topic to the academic community represented by the journal.

shown as sub-components in Figure 4.1, with generalized statements of each subclaim and the type of support most frequently associated with it.

Support for this component comes primarily from past research. The easiest way to support importance to the field is to cite related research articles. If many researchers are working on a topic, that alone establishes its importance. For a new or marginalized topic, the author will need to present stronger reasoning. If something is well known or unquestioned in the field, such as general knowledge about how the research addresses a societal need, citations may not be necessary. We will talk more about how to support claims in Chapters 7–9.

> In engineering, importance may be assumed for recognized problems in the field, so there may be only one sentence on it, or none.

In science articles, importance often constitutes a large part of the Introduction. Since engineering seeks to solve a recognized problem, less time is spent on this component—rarely more than one paragraph and frequently only one or two sentences. Brief articles announcing the latest news on hot topics at the forefront of research or those in specialized journals may not discuss importance at all. This is another difference between IMRD and IPTC formats.

The following examples show how component markers can help identify this component.[7] General importance can be shown by phrases indicating the ability to help solve a recognized problem, the value to society, or frequency of use, as indicated by the first three examples, respectively. Importance to the research field can be demonstrated by the amount of related work, either past or present, as illustrated by the last two. In engineering these may be followed by a citation with a long list of related articles, but in science it is common to mention the contribution of each article separately, increasing the length of this component.

- The rising incidence of . . . has emerged as a great concern . . .
- . . . is of particular value for emerging mobile applications.
- . . . is widely used in . . .
- . . . has been actively studied for decades.
- . . . is increasingly recognized as an important research area for current and future devices.

4.5 Component 2: need

> To demonstrate need, the author must show that the research is related to but different from previous work—that it fills a research gap.

The unclaimed space in your field that your research will fill is sometimes called the research gap (Swales, 1990). The term research gap almost never appears in an article, but is nonetheless a useful descriptive term. In English writing it is important to specify both the research gap and the research goal—what is missing from previous work and how you propose to fill that gap. The term gap highlights the idea that your research cannot be identical to what has been done before, but must be related to it, filling a gap. Showing there is a gap is the purpose of the second component, need for the work.

The importance of the gap can perhaps be best understood by means of illustration. For six years after we moved to the university where I now teach, Google Maps satellite view showed an empty field on the street across from the main gate of the university. I knew it was not empty because I had eaten at several restaurants in that block, which had been developed into a row of shops. From the outdated satellite view it seemed it would be possible to build something on that block, but locals knew otherwise. That space was already taken. The same thing can happen in research. From a distance, a newcomer might think there is unclaimed space in a research field, but those in the area may know otherwise.

In science it is common to identify the gap in previous research, but in engineering this is frequently replaced by limitations (2b), sometimes called problem identification, as shown in Figure 4.2. The distinction is important, because a research gap implies that the particular question posed as the research goal has never been answered, although there may be research on closely related questions. Limitations imply that work has been done on that problem, but the solution is still insufficient. In science a similar situation could exist if the author seeks to show that previous answers are incorrect or inconclusive.

2) Need

a. Research gap (IMRD)
   Claim: There is a gap in current knowledge or understanding
   Support: Absence of past research (nothing has been done)
b. Limitations (IPTC)
   Claim: The current best solution is limited or less than ideal
   Support: Past research (present best solution, limitations)
c. Benefit
   Claim: This will be beneficial to society or the field
   Support: General knowledge or past research

Figure 4.2 Common sub-components of Component 2: need.

In some fields, particularly those with multiple approaches to similar questions, there may be a second section in the Introduction division devoted primarily to need. This is standard practice in computer science and related fields, where the section is frequently called Background or Related Work.[8] Previous approaches are described, along with their limitations. The last paragraph of the section may restate the research goal and how it addresses the need.

It is also possible to claim there is a need based on the benefit of the research (2c). This can be done in both science and engineering. This subcomponent is closely related to the importance of the research and is used where there is an immediate practical benefit to society.

As with Component 1, the main support for Component 2 comes from past research. With a research gap, the *absence* of past research shows its existence, while citations of similar research show the importance of filling it and thus are more closely associated with the first component. On the contrary, citations in engineering more commonly show the *limitations* of past research and therefore are directly associated with the second component.[9]

In engineering, sentences indicating limitations are often found at the end of a long paragraph describing previous research.

Component markers can also help identify the need for the current work. Current approaches are very field-specific but are generally easy to find, as they are usually accompanied by a citation. Sometimes the need is identified as a gap in the research that has not yet been investigated, as in the first three examples below. This is more common in science than engineering. In engineering it is more common to note a limitation, either by means of a contrast (however, although, nevertheless) or terms that indicate disadvantages of current solutions, as in the other examples. *These markers often occur toward the end of a paragraph describing previous research*, making them harder to find.

- So far, nothing has been done . . .
- What has been missing until now . . .
- . . . have not been characterized to date.
- However, a key problem with most algorithms . . .
- Nevertheless, both solutions incur . . .
- These drawbacks . . .
- Although this . . . has favorable properties, it does have some disadvantages, such as . . .
- The two primary concerns are . . .
- The standard . . . are inadequate . . .
- . . . limit the applicability of these methods to broader applications.
- . . . require a series of complicated steps and special instruments. They are high cost, labor intensive, and time consuming.

4.6 Component 3: research goal [main claim]

> The research goal is often found in the last paragraph of the Introduction, or the previous paragraph if the last describes the organization of the article.

Although listed as the third component based on its typical location in a research article, the research goal is the overall claim of the whole article. It is so vital to both editors and potential readers that similar statements appear at least three times. The main claim is usually found near the end of the Introduction section (not division, if there are two sections). It also appears in the Abstract, while the Conclusion (Component 10) says that the goal has been accomplished.

Recall that science usually asks research questions and seeks explanations, whereas engineering defines a problem and seeks a solution. Thus science may phrase the goal as questions while engineering speaks of aims, but many different terms are used. In engineering it is also common to state the proposed solution or the contribution directly, as shown in Figure 4.3. In this case the wording of the research goal and Conclusion may be very similar. Note that, unlike the other components, *no support is listed for the research goal, because all the other components support this main claim.*

3) Research goal [main claim]

a. Specific topic (goal, aim, objective, focus, question) (IMRD or IPTC)
   Claim: We add to previous research in this area
b. Proposed solution (IPTC)
   Claim: We provide a better solution to this problem
c. Contribution (IPTC)
   Claim: We make an important contribution in this paper

Figure 4.3 Common sub-components of Component 3: research goal.

If stated as an objective, the research goal may be a single sentence. For example, in the IMRD article, "Effects of rainfall" (Kleinheinz et al., 2009), the last sentence of the Introduction says (emphasis added):

- "The overall objective of this project was to determine what impact rainfall had on the *E. coli* concentrations at the selected beaches."

In IPTC format, the objective is frequently followed by a brief summary of the proposed solution, and perhaps a statement that as far as the authors know, no one has yet published that proposed solution. All three of these elements appear in the penultimate paragraph of the Introduction to the article "Spatial query integrity" (Hu et al., 2013) as shown by the underlined words:

The research goal may be stated in one sentence, one sentence with numbered or bullet points, or several sentences.

- "This paper subsumes our earlier work in [10] by extending . . . to handle more advanced spatial queries, . . .
- Specifically, we propose . . .
- To the best of our knowledge, this is the first paper that tackles the query verification problem for these types of queries."

Many papers clearly and succinctly state the research goal, using easily recognizable component markers. With others a problem definition may be stated as part of the larger summary of the current state of knowledge in the field, followed by a proposed solution. Nevertheless, in almost all cases these statements occur near the end of the Introduction section and there is some sort of marker to indicate the transition from past to present research, usually indicated by "we," "our," or "this paper," as shown by the underlined phrases below. Sometimes multiple goals are made more conspicuous by using numbered or bullet points.

- This paper tests the hypothesis . . .
- This paper seeks to answer the following questions: . . . [three numbered questions]
- This paper proposes . . .
- Toward this end, this paper presents a methodology . . .
- In this paper we discuss the implementation of . . .
- In this paper, we study the following two goals . . . [two goals listed in sentences]
- In contrast to prior studies . . . here we investigate . . .
- We seek to answer the question . . .
- Here we investigate this issue using calculations . . . We in fact find . . . and provide a plausible explanation for . . .
- Here we examined three specific predictions . . . [three predictions listed in sentences]
- We intend to demonstrate . . .
- In addition, we extended the technique to estimate . . .
- Our goal is to recognize . . . Our system . . . manipulates . . .
- The contributions in this paper are threefold. [three sentences]

- The unique contributions made in this paper are as follows. [four numbered points]
- Our major contributions in this paper thus include: [three bullet points]
- We designed . . .
- We fabricated and tested . . .
- To our knowledge, this is the first . . .

> A research goal hidden in the middle of a paragraph is harder to find, and occasionally is not marked by these expected component markers.

> The research goal states the goal of the article, which may be different from the goal a researcher had in mind when beginning the research.

In a well-written article the research goal will be easy to find, often in a topic sentence at the beginning of a paragraph, but unfortunately not all research articles are well written. Sometimes it is hidden in the middle of a paragraph, following a statement of the limitations of other research, particularly in Asian writing. If you cannot find it, try to identify the goal in the Title, Abstract, or Conclusion, then look for a similar idea in the Introduction. In some IPTC papers it will appear at the beginning of the Process division. Occasionally it will be restated at the beginning of the third division, for reasons discussed in section 6.2.

If you write up your own research it is important to remember that the research goal is the goal of that article, not the overall or original goal of your research. When pursuing research, you will have a goal in mind. When you write it up, you may find that goal has shifted slightly. Rather than redoing the research or rewriting the paper, it is more reasonable to rewrite the research goal. We will return to this idea in Chapter 23 when discussing the final stage of revision.

4.7 Component 4: framework

As noted above, research is almost always related to an established framework.[10] If a certain framework guides most of the current research in a field or journal, it may not be specified in the article. If there are multiple ways of approaching a question, or if a researcher uses a non-standard approach, it is much more important to explicitly state the framework.

> The framework may be found in either the first or second division, and is often inferred rather than stated directly.

In science the framework often can be inferred from the author's attitude toward past research in the Introduction and the choice of Methods. In engineering the framework may be the last method mentioned in the Introduction, with the fewest limitations, then discussed more extensively as the foundation of the research in the Process division. Since in both formats the framework may be identified from both the first and second division, we will leave a complete explanation and analysis of it until the next chapter.

Teaching tip: Covering Component 4 with the second division balances the work better and is less confusing for students, since it is logically connected with the second implicit claim.

4.8 Structure of a well-written article

Before you analyze your exemplar articles, let's look at an example of a well-written engineering article, "Spatial query integrity" (Hu et al., 2013). The first sentence of each paragraph is a topic sentence that identifies the purpose of that paragraph. When you do your own writing, I encourage you to use topic sentences, as this makes it far easier for you to frame your argument and for your readers to find the information they need. In Table 4.1 the main component of each paragraph in the Introduction has been identified.

Note that the analysis is not simply based on keywords. For example, paragraph 4 mentions the word "framework," but talks about the framework most common in *other* research, not this article, and therefore has been assigned to Component 2, since it gives information about the current best solution and its limitations.

Note also that the last paragraph in the Introduction section is labelled Organization. This is rare in science, but common in both engineering and social science. Its purpose is to help the reader follow the logic of the argument, as will be discussed in Chapter 9.

Table 4.1 Components in the first section (Introduction) of "Spatial query integrity"

Paragraph	*Topic sentence with component marker*	*Component (purpose)*
1a	The amount of digital spatial information available . . . has grown at an exceptional pace over the past decade.	1a (General importance)
1b	Consequently, . . . is becoming increasingly popular and has received a lot of attention in the research community.*	1b (Specific importance)
2	In this work, we focus on the . . . model, . . . However, there exist two major concerns with this model.	2b (Limitations)
3	The general framework commonly used in the literature . . . is based on . . .	2b (Current best solution)
4	The current state-of-the-art solution . . .	2b (Current best solution)
5	We argue that . . . suffer from several drawbacks.	2b (Limitations)
6	Motivated by the above observations, we propose . . ., a novel approach . . . based on . . .	3b (Proposed solution)
7	This paper subsumes our earlier work . . . by extending the . . . approach . . .	3a (Specific objective)
8	The remainder of the paper is organized as follows.	[Organization]

*This is actually the second part of the first paragraph, which discusses both levels of importance, but in other articles the two aspects may be in separate paragraphs.

Table 4.2 Components in the second section (Related Work) of "Spatial query integrity"

Paragraph	*First and last sentences of each paragraph, with component markers*	*Component (Purpose)*
1	The idea . . . was first introduced by . . . Nevertheless, both solutions incur significant authentication overhead . . .	2b (Limitations)
2	. . . introduced . . . supporting fast query processing and verification. Nevertheless, neither . . . handle data updates efficiently.	2b (Limitations)
3	Efficient verification in the presence of frequent updates has been studied . . . The proposed mechanism achieves both correctness and . . .	2b (Current best solution)
4	All the aforementioned solutions require changes . . . The . . . method introduced in this paper is both efficient and deterministic, and does not require any modifications . . .	2b (Limitations) 3c (Contribution)

Table 4.3 Introduction division of the sample article "Spatial query integrity"

Section	*Paragraph*	*Components*
1	1a	1a Importance (general)
	1b	1b Importance (specific)
	2–5	2b Need (limitations)
	6	3b Research goal (solution)
	7	3a Research goal (objective)
	8	Organization
2	1–4	2b Need (limitations)
	4	3c Research goal (contribution)

In IPTC, the Introduction division may include a more extensive Related Work section, as is the case with this article. There are only six citations or strings of citations in the Introduction section of this paper, but 15 in the second section, in half as many paragraphs.[11] As shown in Table 4.2, the focus of this section is limitations (2b). Each of the first two paragraphs state the benefits of a method but end in "Nevertheless." While the third paragraph accentuates the value of previous methods, the final paragraph again points to their limitation and ends with a restatement of how the present work overcomes this limitation.

Thus, the structure of the Introduction division of this sample article could be summarized as shown in Table 4.3. Eight of the twelve paragraphs and all but two of the citations are associated with need.

4.9 Comparison of IMRD and IPTC

To conclude, Components 1–3 are found in the Introduction of almost all research articles, usually in order. Component 4, the framework of the research, may be identified from either the first or second division, or both, often by the attitude of the researcher toward previous research rather than a direct statement.

> There is considerable variation in the structure of the Introduction in IPTC, in length, components present, number of citations, and presence of graphics.

There is far more variation in the Introduction in IPTC than in IMRD. Sometimes there are no citations at all, either because the Introduction is very short, only one or two paragraphs, or because most of the citations are in a Related Work section, which is rare in science. In engineering the Introduction may also be very long, sometimes with a figure or table showing an overall schematic of the design presented in the paper, while graphics are rare in the Introduction to science articles.

As mentioned above, importance (1) tends to be more prominent in science, but need (2) in engineering. In engineering, different aspects of the research goal (3) may be stated in separate paragraphs, with each paragraph discussing one characteristic that needs to be incorporated into the new design, followed by that part of the goal. If stated as a contribution, the wording of the research goal may be very similar to the Conclusion. Furthermore, engineering articles frequently conclude the first section with a paragraph showing the organization of the article, which is rare in science. Some of these differences may be due to less editing of engineering articles, but others appear to be due to the innate differences between science and engineering research.

Exercise 4.1 Introduction division components

(1) Identify the components in each paragraph in the Introduction division of your first exemplar article and the component markers that help you identify them. Summarize this in a table.

Hint: Use the list of components in Appendix 1 to avoid flipping pages.

Hint: Remember that component markers are general words that could help identify that component in many different articles, not the specifics of that article.

Hint: Mark what components are present in each paragraph in the margins of your article. After collecting the data, decide how to summarize and present them. Look at the tables in this chapter and the online supplemental materials for ideas, but do not feel bound by them.

Hint: If your article contains Related Work or a similar section, remember to include it in the Introduction division.

Hint: If you are in a writing class, your instructor may ask you to share with the class examples of component markers. If not, consider doing this with a friend. This will help you learn to recognize the structure of the articles more quickly.

Hint: Past research may be cited to support the importance, need, or framework. We will look more at past research in Chapter 7.

Teaching tip: Direct students' attention to the online supplemental materials, which include a complete component analysis of one article each in science and engineering.

Teaching tip: Asking students to mark the components on the article (hard copy or electronic) eliminates questions about what they consider a paragraph, which is sometimes difficult to determine if there are many equations or other non-text items, common in engineering articles.

Teaching tip: Students may list specific content rather than general components. Help them to identify the general components rather than the specific content of the article, which is probably all they have focused on previously.

(2) Does the Introduction follow the order of components as listed?

Hint: Practice writing your answers as sentences that include the question, rather than using a question-and-answer format. For example, you could say, "In this article the order of components in the Introduction fits the expectations for IMRD."

(3) Where is the research goal stated in your exemplar articles, and how? If they use the phrase "to [the best of] our knowledge"[12] or additional bullet points or sentences to expand on the goal, make note of that.

Hint: If you do not find the research goal in the Introduction, it is occasionally found in the first paragraph of the Process division in IPTC format.

Teaching tip: You might ask students to identify the research goal before the other components commonly found in the Introduction. Since it is important and usually easy to recognize, it provides a good starting point.

(4) Is there a paragraph at the end of the Introduction that describes the organization of the rest of the paper?

Hint: This is common in IPTC, but not in IMRD.

Teaching tip: If this is too many questions for the level of your students, questions 1–2 are the core questions.

Evaluate your understanding

(1) Why is it important to be part of an academic, disciplinary, or research community? How does the Introduction help you claim your role in that community?
(2) What common components are generally present in an Introduction, and what is the role of each in supporting the main claim?
(3) Why is the research goal the most important component of any research article?
(4) How do topic sentences and component markers help you, both as a reader and a writer?
(5) How do Introductions differ in IMRD and IPTC format?

Homework

(1) Repeat Exercise 4.1 for your other two exemplar articles.
(2) After gathering all your data, summarize them in the form of a claim and support. Note anything special about individual articles, for example paragraphs on organization or the location of the research goal.

Hint: For example, "My first two articles follow the expected order of components, with a final paragraph on organization, as shown in Table 2a and 2b, but the third has an unusual structure. As shown in Table 2c, . . ."

Hint: Add this to what you wrote in the last chapter to build a more complete picture of the general expected structure of articles in your field.

Teaching tip: It may be useful to set up a system of peer review. This has multiple advantages:

1 *As a writer, peer review gives students additional feedback.*
2 *As a reviewer, peer review develops the skill of reviewing others' work.*
3 *Peer review instills the importance of collaboration in academic work.*
4 *Students will often notice things that writing teachers may not, or the process may reveal misunderstandings or questions that the students may be unwilling to ask in front of the whole class.*
5 *For students using English as a lingua franca, it also gives an opportunity to practice spoken English and listen to different Englishes.*
6 *Peer review allows students to see more examples of how similar components are expressed in other fields, as well as different organizational structures.*
7 *Students are actively engaged in the material and the revision process, and always report the peer discussion is an important and enjoyable part of their learning.*
8 *In the process, students sometimes make new friends. Switching the groups every week or two maximizes these advantages.*

Teaching tip: Consider teaching the chapters on support at the same time as the chapters on divisions, especially for advanced writing classes: Introduction (4) with Evidence from Past Research (7), Methods/Process (5) with Evidence from Current Research (8), and the final two divisions (6) with Reasoning (9).

Supplemental material: Sample component analysis.

Notes

1 Each framework has its own set of presuppositions, often about questions outside the realm of science such as human nature, desired results, or ethical standards (see section 3.1).

2 Many linguists have proposed a similar structure for the Introduction division. For example, Swales' CARS model (1990: 141) and revised CARS model (2004: 230, 232) propose three moves that parallel my three components, but Swales associates past research only with Move 1. This is usually true in science but less so in engineering. Likewise, the 6 Stages listed by Cargill and O'Connor (2013: 44) for the Introduction include my components 1–3 (splitting research goal into two stages, goal and contribution), but also include past research and organization, which I consider evidence and reasoning, respectively. Other similar models include Lewin, Fine, and Young (2001) and the OARO model (Swales, 2004: 244). My component list was generated independently, modified in response to student feedback, and largely completed before I saw any of these, supporting the categories identified by previous authors, but I have not followed any of their naming systems for several reasons. First, none have related the components to argument structure, to an implicit general claim. Second, none have separated the three component claims

from evidence, in this case past research, nor have they noted the distinction between science and engineering in the primary location of citations of past research. Third, the names they have used (moves and steps or stages) suggest to non-linguists that the components follow a certain order, which is usually true for the Introduction, but not necessarily for the other divisions. More details will be presented in Rau (In preparation).

3 Many linguists have split paragraphs, and even sentences, in their analysis of moves or stages. Analysis by paragraph uses a clearly identifiable feature to attempt to reconstruct the intent of the author.

4 This is definitely true in Chinese, and based on feedback from my students appears to be a common feature of many Asian languages.

5 "Lexical bundles" are common in engineering writing and can be used effectively to teach the rhetorical conventions of a discipline (Hyland, 2008). The same concept is also called "sentence patterns" (Cargill & O'Connor, 2013).

6 Since component markers are portions of sentences that are common and not content-specific, this is not considered plagiarism.

7 Throughout the book, the component markers listed are taken from published articles, with words specific to that article removed and the remaining clause edited for clarity if necessary.

8 Related work is very similar to the Literature Review often found as the second section in social science articles.

9 "Reviewing items of past research" is Move 1, Step 3 of the revised CARS model (Swales, 2004: 230, 232). In engineering this is more frequently associated with Move 2, although since the review precedes the claim (e.g. "nevertheless"), the order is the same. See Rau (In preparation).

10 Even theoretical work to establish a new framework must show how it improves on the current one.

11 Citations can be counted either by number of occurrences or number of articles cited. The former counts the number of ideas that need to be cited, ignoring the number of articles cited in each string and multiple citations of each article, and is easier practically. The latter counts each citation in a string separately but eliminates multiple citations of the same source.

12 "To the best of our knowledge" is considered proper use of the set phrase in American English, but "to our knowledge" is increasingly appearing in articles in the ELF world.

References

CARGILL, M. & O'CONNOR, P. 2013. *Writing scientific research articles: Strategy and steps*, New York, John Wiley & Sons.

HU, L., KU, W.-S., BAKIRAS, S. & SHAHABI, C. 2013. Spatial query integrity with Voronoi neighbors. *IEEE Transactions on Knowledge and Data Engineering*, 25, 863–876.

HYLAND, K. 2008. As can be seen: Lexical bundles and disciplinary variation. *English for Specific Purposes*, 27, 4–21.

KLEINHEINZ, G. T., MCDERMOTT, C. M., HUGHES, S. & BROWN, A. 2009. Effects of rainfall on *E. coli* concentrations at Door County, Wisconsin beaches. *International Journal of Microbiology*, https://doi.org/10.1155/2009/876050.

LEWIN, B. A., FINE, J. & YOUNG, L. 2001. *Expository discourse: A genre-based approach to social science research texts*, London, Continuum.

RAU, G. In preparation. IPTC: A prototypical format for engineering research articles.

SWALES, J. M. 1990. *Genre analysis: English in academic and research settings*, Cambridge, Cambridge University Press.

SWALES, J. M. 2004. *Research genres: Exploration and analysis*, Cambridge, Cambridge University Press.

Chapter 5

Demonstrating the continuity and novelty of your research

5.1 Connected yet distinct

In some communities every house or apartment on a street is identical in construction. Yet as soon as someone moves into one of the units, they begin to transform it to reflect their interests. One person might put a flowerpot outside the door, another hang a flag or other ornamentation. Inside the house the differences will be even greater. The same is true of a research article. On the surface, the second division of many papers will appear very similar, particularly in science, but as you look at the details, you'll find each has its own distinctive features.

On the other hand, houses around the world may be very different in structure, but all fulfill the same purpose of providing shelter and all have rooms or areas that serve the functions of living, dining, and sleeping. Similarly, the purposes of the Methods division of IMRD and the Process division of IPTC are quite similar, despite major differences in structure.

> The second division does not merely report what you did, but more importantly how it is both connected to yet distinct from previous research.

The second division is not merely a description of what you did. Like the Introduction, it serves the far more important function of placing your work in the context of what has already been done. Much engineering research involves minimal changes to previous research, just as in science one variable is changed at a time and the results measured. Thus the Methods section in IMRD is not merely an instruction manual, but shows how your work is similar to previous research while highlighting the differences. Similarly, the Process division in IPTC is a demonstration that you have designed something new that improves on previous work.

5.2 Claim 2: reliability or feasibility of your work

Most of the second division is devoted to details of the current method or design. Recall the claim the author is trying to support in this division:

2 You have gathered data in a valid, reliable way (IMRD)

or

2 You have designed a workable solution to the problem (IPTC).

This claim is established by providing details of the present research, supported as needed by references to past research. In IMRD, the details allow others to validate the method, and your attention to detail in the description vouches for similar attention to detail in carrying out the procedure and thus the reliability of the data. In IPTC, the attention to detail allows others to verify your solution is workable. Yet the purpose of presenting the detail is quite different, because the focus of IMRD is the data, which will be presented in the following division, whereas the focus in IPTC is the new design described in this division.

This difference in focus is reflected in very different division structure. In IMRD the Methods division may be very similar to previous articles, so it tends to be written in a very concise style and may even be placed at the end of the article in smaller font. In contrast, in many IPTC articles the Process division is the longest division and written in an extended style, with very little similarity between papers.[1]

Since the present research builds on past research, it is quite common in both formats to find Component 4: framework in this division, but part of the extra length in IPTC may be due to a long initial description of the framework. In both formats much of the division will be devoted to Component 5: research details. Because of the difference in the implicit claims, Component 6: testing methods is usually found in the Methods division in science, but the Testing division in engineering. Statistical testing and the related experimental design are far more important in science than engineering, as will be discussed at length in section 8.3.

5.3 Component 4: framework

Many terms could be used for this component, including framework, approach, model, and theory. While these are not interchangeable, definitions given for them overlap. As I will use the terms, framework is a general way of addressing a topic. Approach is often more specific, referring to tools or methods developed by previous authors. Model refers to a theoretical, conceptual, or mathematical basis of the research. Widely used models with broad explanatory ability are called theories. For the purposes of this book we will use the general term framework, but your field may prefer another term.

> Previous research can be cited to reveal its limitations (need) or positive aspects the current work builds on (framework).

Framework provides background information about previous research to show how it lays the foundation for the current research. Framework (4) can be distinguished from need (2), because *need focuses on the limitations of previous research, whereas framework focuses on the advantages of one method.*

Experienced researchers in a field will be able to identify the framework based on the author's attitude toward past research in the Introduction. Since people talk positively about things important to them, more time and positive words will be devoted to talking about research using the same framework. In science the framework can often be inferred based on what the author thinks important and which previous research is evaluated most positively, or the way the research goal is stated.[2] In engineering the framework may be indicated by the last of the previous methods presented, with the fewest limitations.

The framework is often also indicated in the second division, by the choice of methods followed or the foundation of the present design. The framework can vary in length from a single sentence to a whole section. Some of the ways it may appear are listed in Figure 5.1.

In IMRD articles the framework can often be inferred from the methods used (4a), since each methodology tends to be based on a certain approach or theoretical model. Explicit discussion of the framework (4c) is less common in science than either social science or engineering, but if different methods could have been used there may be a one paragraph justification (4d) at the beginning of the Methods.

In IPTC, a subsection entitled problem identification or problem formulation may be found in either the first or second division. When found in the Introduction, the focus is usually on the limitations of previous research (2b). On the contrary, in the Process division the focus is normally on positive aspects of a previous design that the present work builds on (4b). There may be a section on the mathematical basis of the current work (4c), including definitions and nomenclature. There may also be a comparison of previous approaches, and justification of the one chosen (4d). For a very new solution, there may be little or no framework.

The following component markers indicating the framework were all closely connected with the research goal in the Introduction, either in the paragraph before, earlier

4) Framework

a. Accepted procedures (IMRD)
   Claim: Accepted procedures or proven methods were followed
   Support: Past research (citations or general knowledge)
b. Problem formulation (IPTC)
   Claim: The research extends or builds on an established approach
   Support: Past research (advantages of previous research approach)
c. Model or theoretical framework
   Claim: The research is based on an accepted model or established design
   Support: Past research (model, framework, mathematical basis, design)
d. Justification
   Claim: This is the best method to follow for our research
   Support: Past research (alternative methods)

Figure 5.1 Common sub-components of Component 4: framework.

in the same paragraph, or part of it. The keywords underlined either highlight the benefit of a method or how the current research builds on previous work.

- The excellent fluorescent properties give utility for . . .
- The possibility of incorporating . . . makes the sensor a potential candidate for . . .
- The proposed solution relies on . . . a useful paradigm to solve . . .
- . . . exhibits exceptional features useful in diagnostic applications.
- In this paper we extend the approach . . .
- In this work, a novel regularization technique is introduced that builds on . . .

If there is a second section in the Introduction division, some paragraphs may be designated as framework if they focus on how the current method builds on the *strengths* of previous methods rather than overcomes the *weaknesses* of past work. This seems to be relatively uncommon.

> **Framework presents background information needed to understand the present study, rather than details of the present study, but sometimes the border is unclear.**

In IPTC the framework may comprise a whole section at the beginning of the Process division or the first subsection of several sections. Rather than merely referring readers to previous literature, it is common in engineering to summarize either the theoretical principles underlying the present design or ideas borrowed from previous designs. This makes it easier for the reader to understand and evaluate the current design details, presented in the following paragraphs or section, without looking at previous articles. The framework is distinguished from the research details because it is a description of background information or previous work needed to understand the current study, rather than the new design. Sometimes the boundary between the two is fuzzy, particularly if the current research is an extension of previous work.

Sometimes entire sections or subsections can be identified as framework based on the section titles, for example:

- Background information on . . .
- Components of [the system used in previous papers]

Since the framework is very field-specific, it is hard to find general component markers. Nonetheless, the following talk about specific procedures that would be classified as 4a.

- In order to understand . . . we briefly explain . . .
- Our target architecture is . . . , the basic idea of which is . . .
- In this paper, the method is transformed . . .
- In . . ., a fast method to measure . . . was proposed. . . . This method also can be extended to measure . . .

In IPTC the framework sometimes includes problem formulation (4b), often as a separate section. This is common in articles developing a new algorithm or mathematical procedure, although the connection with previous work is often stated in the previous section.

- Based on the aforementioned placement and routing models, . . . we define the problem as follows.
- With the cost definition, we formally define the addressed problem as follows.
- In this section, the algorithm given in Section II is extended to the problem . . .

The following are examples of 4c, because they explicitly refer to a model or theoretical framework.

- As shown in Fig. 1, the . . . model represents . . .
- According to the analysis of motor theory . . .
- This section is devoted to a brief review of the theory . . .
- Our proposed setup can be placed in the category referred to as . . .

> Justification (4d) answers the question "why" this was chosen while the other sub-components answer the question "how."

Justification of the choice of a framework (4d) answers the question why, whereas the other sub-components answer the question how. Since it involves a comparison and choice between two approaches, general component markers can be identified more easily. Justification can take the form of purpose (in order to), cause or result (since, consequently), avoidance or reduction of problems, or need for correct measurement or interpretation, as illustrated by the following examples:

- In order to take advantage of . . . and lower the deployment cost, . . .
- Since this device is meant for . . ., this frequency is a good choice.
- Consequently, . . . is required.
- Thus, using . . . as calibration parameter to correct . . . could facilitate accurate . . .
- . . . [problems] can be theoretically avoided.
- The motivation for the . . . concept is to reduce . . .
- To interpret the data correctly, . . .
- Selecting the most appropriate technique is governed by . . .

This component is supported by previous research but may not have citations. Subsections dealing with definitions, mathematical notation, or well-known basic information may be general knowledge and widely used by many in the field. More recent work specific to the current research will probably have citations.

Sometimes framework and research details will alternate through the Process division, with background information on one concept, then details of the current design that build on that concept, followed by a second concept and its related design. This happens frequently in papers that describe several different design elements in a single paper, for example hardware, software, and an algorithm that work together to accomplish a single research goal, and in papers that combine previously existing methods.

To differentiate framework from other components, keep the following questions in mind:

1 Is this describing limitations of previous research (2) or advantages the current work builds on (4)?
2 Is this background necessary to understand the current research (4) or the current research itself (5)?

To return to the metaphor of a house in a community, remember that need (2) talks about existing houses, framework (4) lays the foundation, which will be very similar to others but specific to the present work, and research details (5) are the things that clearly separate the new structure from others.

5.4 Component 5: research details

Since there are many different research methodologies, Component 5 is extremely variable. A generalized list of sub-components common in science and engineering is shown in Figure 5.2.[3] Do not feel constrained by the sub-components suggested. It may be useful to add categories for things specific to your field. For example, in chemical engineering there are frequently subsections on fabrication and characterization, so these would be useful terms. A different list would be required for social science. For example, social science often has sections on participants and ethics and may include an extensive description of qualitative methods. Some areas of science and engineering that involve human subjects, such as ergonomics, use a reporting format similar to social science and include these topics.

Different types of research details tend to be presented in science and engineering. In IMRD, they are commonly restricted to the first two sub-components. Note that I have

5) Research details
a. Materials obtained
Claim: Appropriate materials were obtained from reputable suppliers
Support: Names and sources of materials
b. Methods followed
Claim: Care was taken to ensure good results
Support: Details of present research design and procedures
c. New design
Claim: A workable new solution, product, or process was developed
Support: Details of new design, intermediate testing of parts
d. New mathematical procedures
Claim: A mathematical theorem, procedure, or algorithm was developed
Support: Details of proof, lemmas, mathematical argument

Figure 5.2 Common sub-components of Component 5: research details.

chosen to call these Materials *obtained* and Methods *followed*. This may not be necessary in science, but in engineering it helps to distinguish Materials and Methods (5a and 5b) developed previously by others from those developed by the current authors to fulfill the goal of the present work (5c and 5d). I have chosen not to separate Product from Process as sub-components because sometimes it is hard to distinguish the two, for example in the development of a new circuit design or software, where a process is built into the new product.

IMRD may not clearly specify which parts of the Materials and Methods follow previous research and which are unique to the present work. While there may be some modification of previously used methods, the focus of science is on the data to be collected rather than the novelty of the method.[4]

On the other hand, in IPTC it is far more important to distinguish past from present in this division. In engineering some materials will be purchased, and some methods followed that were used in previous work, but the goal is development of a novel product or process, so the third sub-component will dominate. In some fields of engineering new mathematical procedures comprise a large portion of the Process division, so this is listed as a separate sub-component. If the mathematical procedure is the main goal, the paper may lack a Testing division, like other mathematical articles. In some fields of engineering the design process is reported in stages, rather than just the final design. In this case there may be intermediate testing between each design modification. Thus Components 5 and 6 may alternate within the Process division, again without a separate Testing division.

To identify the sub-components, it may be sufficient to examine the section or subsection titles, if they clearly designate the authors' intent. For example, in the following section titles from an article in chemical engineering, section 2.1 refers to materials (5a), whereas section 2.2 describes methods (5b), as evident from the word "production." Section 2.3 then describes testing, Component 6, explained below.

2 Experimental

- 2.1 Materials
  - 2.1.1 Raw materials
  - 2.1.2 Characterization
- 2.2 Activated carbon production
- 2.3 Adsorption tests

> Sub-components 5a and 5b are obtained from others and important in IMRD, but 5c and 5d are developed in the current research and more important in IPTC.

Beyond section titles, it is helpful to look for component markers, especially in IPTC format. Unfortunately, since the materials and methods used or the process and mathematics developed are extremely field-specific, it is hard to list general examples. As you develop a list of markers used in your field, keep in mind that 5a and 5b are things obtained from others or from previous research, whereas 5c and 5d are items developed and described for the first time in that paper.

5.5 Component 6: testing methods

Another important difference between IMRD and IPTC is the location of the testing methods, Component 6 (Figure 5.3). In IMRD the testing methods are usually included in the Methods section, often as a separate subsection (as in the example above), whereas in IPTC they are more commonly described in the Testing division, frequently in the same section, or even paragraph, that reports the results of the test. If the design process is reported as a series of steps, rather than simply presenting the final design, testing or optimization will follow each design step to determine which parameters to use in the next step, so Components 5 and 6 will alternate, and there may not be a separate Testing division.

In IMRD testing methods are usually described in the Methods division, but in IPTC they are often found in the Testing division, just before the test results.

There is a further difference in the types of testing used in the two formats. While physical tests are

Science usually employs physical testing, often with statistics, whereas computer or simulation testing is more common in some engineering fields.

6) Testing methods

a. Physical testing
   Claim: Testing was done carefully, following verifiable procedures
   Support: Details of data collection method
b. Computer testing (IPTC)
   Claim: Testing was done on a standard data set
   Support: Details of testing method
c. Simulation testing (IPTC)
   Claim: Testing was done under realistic simulated conditions
   Support: Details of testing method
d. Statistical testing (IMRD)
   Claim: Testing was done using standard statistical tests and tools
   Support: Names of tests and software used, details of statistical design

Figure 5.3 Common sub-components of Component 6: testing methods.

common in both, computer and simulation testing are rare in IMRD. Physical testing involves measurement of data from physical objects. Computer testing uses electronic data sets derived from real conditions. Simulation testing uses idealized data created by the researcher. For example, in a robotic vision lab, physical testing could be done with cameras and physical objects, computer testing could be done using 3D computerized images of real objects, or simulation testing could be done on idealized grid patterns. Tests done on a computer without mention of a data set should normally be classified as simulation.

Statistical testing may be used in conjunction with any of these three testing methods. It is very common in science but relatively rare in engineering. When statistics are used, it is essential to include a careful description. Far too many engineering papers that include error bars in the figures contain no information on how those errors were calculated, whether from 5 trials or 1000, which makes a huge difference in how the results are interpreted.

As with Component 5, section and subsection titles may state the type of testing performed. Component markers can also be used to distinguish the various types of testing methods. Often a sentence at the beginning of the description of the testing will identify it, as in the examples below. Similar sentences are sometimes buried in the middle of a section, making it harder for the novice reader, but experienced researchers in a field will readily identify the testing type from the results.

Physical testing (6a), common in both science and engineering, can often be recognized by reference to measurement of physical variables, as in the first two examples below, or testing of physical samples, as in the next two. Development of a prototype can also indicate physical testing. In chemical engineering, success is sometimes measured by whether the desired product was obtained, so testing may consist of characterization of that product, as in the last two examples.

- We measured the angle of the diffracted beam . . .
- To test the effectiveness of the proposed approaches, we measured motion . . .
- Different typical industrial pollutants were used to test . . .
- To evaluate the adsorption capacity of the sample, . . .
- The prototype was fabricated . . .
- The end product is characterized for yield and [physical characteristics].
- The surface morphology of the sample was characterized using SEM.

Computer testing (6b) uses real data sets, but the testing is done and measurements obtained on a computer. Sentences that talk about data sets, databases, or real-world data indicate computer testing. The last example, which occurred in the Testing division, reports three components in the same section: the testing method, results of those tests, and comparison with previous work.

- The . . . digital data set was used . . .
- To evaluate our method, we used three different databases: . . .
- The proposed models have been tested on real-world traffic data.
- This section compares . . . against several state-of-the-art algorithms . . . on the . . . data sets.

Simulation testing (6c) can often be identified by words related to simulation, as in the following:

- We began our testing with a simulation of . . .
- In this section we present simulation results and assess the performance . . .
- We perform simulations to evaluate our proposed method.
- The effectiveness of the proposed distributed control algorithms will be verified by simulating . . .

Other types of testing carried out on a computer that do not use standard data sets would also be considered simulation, including:

- The proposed placement method was implemented in the C++ programming language and was applied in a design example to verify its effectiveness.
- In this section we demonstrate the use of the normal vector method . . . by applying it to . . .
- To evaluate the computational efficiency of different methods, we study . . .
- Experiments have been performed to compare the proposed algorithm with a recently proposed approach.

Statistical testing (6d) can usually be identified by words associated with statistical tests (or their measures), such as mean (standard deviation, standard error, confidence intervals), t-test or χ^2-test (p-value), regression (r^2), or correlation (r). Common with physical testing in science, statistics can be calculated any time there are repeated measures of the same value, including both computer and simulation testing. Often in engineering a graph will show error bars output by a computer program, or a table will show p-values, but there is no indication in the text how the statistical testing was done. Whenever statistical testing is conducted, it is important to report not only the test done, but also the experimental design and predictions, so if any of these terms or features occur in your exemplar articles, pay attention to section 8.3. Examples of component markers that indicate statistical testing include:

- The 95% confidence intervals of the mean . . . are shown . . .
- The level of interference . . . was analyzed using a χ^2-test.
- . . . showed a significant increase in . . . ($p < 0.01$).
- . . . there was a strong correlation between . . . ($r^2 = 0.864$).

Remember that in some cases involving development of mathematical algorithms, no testing is performed. These articles are structured more like articles in applied mathematics, using mathematical proof rather than testing to verify the conclusions, without a Testing division (IPC).

When doing the component analysis, always remember to think about the reason information is included rather than focusing on individual words. For example, you may see the words "result" or "compare" in your article and think those keywords mean that a paragraph should

> Pay attention to the overall purpose of the paragraph rather than individual words. Intermediate results or comparisons may be listed as part of the design process.

be listed as Component 7 or 8, described in the next chapter, but if the verb indicates potential, as in "we can compare," it is probably talking about intermediate testing done as part of the development process. Intermediate testing may also be described as results, "using these results."

5.6 Comparison of IMRD and IPTC

The difference between IMRD and IPTC formats is most obvious in the second division. While two IMRD papers may have nearly identical Methods divisions, there is very little similarity between Process divisions in IPTC. Furthermore, the Methods division is often very short, written in a condensed style, while the Process division is frequently more than half the paper and written in an extended style.

> Organizational features that help the reader, such as topic sentences and introductory paragraphs for each section, are common in the Process division in IPTC.

We will talk more about extended and condensed style in section 9.2, but the most obvious difference is the presence of summary paragraphs and topic sentences in extended style. These features help the reader follow the argument, far more important with the new design presented in IPTC than the relatively familiar method presented in IMRD.

Where present, an introductory paragraph may serve different functions. Often it serves as an overview, providing an organizational structure or summary of a section before giving details. It may also restate the research aim and show how the section contributes to that aim. The only common use of an introductory paragraph in the Methods division of IMRD is to discuss why one approach was followed rather than some alternative (4d).

For example, if we look at the IPTC article "Spatial query integrity" (Hu et al., 2013), we find that most of the research details involve mathematics, but there is an introductory paragraph at the beginning of each section and most subsections explaining the purpose of that procedure. For example, the first paragraph of section 4, Data Transformation, introduces why those calculations are needed: "Before transmitting the database to the SP, the DO transforms each object by . . ." Similarly, section 5 has an introductory paragraph before launching into various subsections relating to various aspects of the authentication process. Most paragraphs have a clear topic sentence.

By contrast, in the IMRD article "Effects of rainfall" (Kleinheinz et al., 2009), there is no introductory paragraph for any subsection in the Methods. Every sentence makes a different point, with no topic sentences and no repetition. This contrasts sharply with the Introduction, which has topic sentences for each paragraph. This is quite common in IMRD papers.

As mentioned above, in IPTC articles framework tends to be far more prominent, since the new design builds on the strength of past designs. Testing is often not mentioned in the Process division, unless the design process is described in stages, with intermediate testing before making the next design decision.

Exercise 5.1 Methods or Process division components

(1) Identify the components in each paragraph in the second division of your first exemplar article and the component markers that help you identify them. Summarize this in a table.

Hint: Remember that it is more important to identify the components than the specific subcomponents, which vary more between articles.

Hint: Both the titles of the subsections and the introductory paragraph (if present) may give hints about what components to expect.

Hint: You may not have to identify every paragraph individually. Look for major transition points, indicated by component markers. Particularly, where is the transition from "they" to "we"? Nevertheless, in your table show how many sections or paragraphs are devoted to each component, as this is an important aspect of the article structure.

Hint: Ask yourself the questions, "Is this background necessary to understand the current research (4), rather than something actually done or developed in the current research? Does this describe something that has been used before, perhaps with modification (5a or 5b), or is developed by the current authors as part of the research goal (5c or 5d)?"

Hint: This division is very field-specific, so if something constantly recurs in your exemplars, for example "fabrication conditions," you may want to use that as a subcategory rather than the more general ones suggested here. Just make sure it is a general category you can use, not specific to the content of one article only, like "fabrication of [specific molecule]."

Hint: Component markers are also field-specific and will be easier to recognize after you have read several articles.

Teaching tip: Students may find it useful to talk with other students in their field to identify component markers.

Teaching tip: Help students determine what is specific to an article, and what are more general recurring categories in their subfield.

(2) Does the second division follow the order of components as listed?
(3) Are there any organizational features such as an introductory paragraph? Are there any other important features to help the reader, such as a figure showing the overall process?

Hint: These are common in engineering, but not science.

Teaching tip: Question 3 is necessary for an advanced class ready to write their own paper, but could be eliminated for a basic class.

Evaluate your understanding

(1) How is the purpose of the second division similar yet different in IMRD and IPTC?
(2) How do the components present in the second division differ in IMRD and IPTC?
(3) Why are organizational features like summaries and topic sentences more important in the second division of IPTC than IMRD?

Homework

(1) Repeat Exercise 5.1 for your other two exemplar articles.
(2) After gathering all your data, summarize them in the form of claims about the general structure of your articles and support for those claims. Note anything special about individual articles that differ from the general pattern.

Hint: Use the format of a journal in your field for tables and table titles.

Hint: If several paragraphs in a section or subsection are the same component, rather than listing each individually you could use a designation such as "Paragraphs 3–9: 5d, Math."

Hint: Consider how to combine the information from different chapters to create a comprehensive picture of the expectations for an article in your field, without repetition of information. Which of the following would be most useful?

1 *Article by article: Article 1 (Overview, Division 1, Division 2); Article 2; Article 3.*

2 *Division by division: Overview (Article 1, 2, 3); Division 1, Division 2.*

Teaching tip: Help the students evaluate the two organizational strategies.

Teaching tip: Particularly in this division, students may be misled by individual words, rather than looking at the context to identify a component.

Supplemental material: Sample component analysis.

Notes

1 Computer science normally follows IPTC format, but builds so closely on prior research that occasionally the first part of the Process is identical to a previously published article, appending a new portion at the end. Identical wording is considered plagiarism in most other fields.
2 The approach used to answer the research question is much more likely to be stated directly in social science.
3 Previous authors have split Materials and Methods into distinct moves or stages, whereas sub-components 5c and 5d are rare in science and thus in previous lists. In naming components, I have sought to identify elements present in most research articles. Since most research articles contain at least one sub-component but few have all, I have grouped them as a single component.
4 When a new method is presented, the article will take on a structure very similar to IPTC. Such papers may be highly cited, but are not representative of most articles in the field.

References

HU, L., KU, W.-S., BAKIRAS, S. & SHAHABI, C. 2013. Spatial query integrity with Voronoi neighbors. *IEEE Transactions on Knowledge and Data Engineering*, 25, 863–876.

KLEINHEINZ, G. T., MCDERMOTT, C. M., HUGHES, S. & BROWN, A. 2009. Effects of rainfall on *E. coli* concentrations at Door County, Wisconsin beaches. *International Journal of Microbiology*, https://doi.org/10.1155/2009/876050.

Chapter 6

Verifying the contribution of your research

6.1 Staking your claim

To stake a claim originally meant to pound wooden stakes into the ground to mark off an area of formerly unclaimed land as your own. The person making the claim had to prove that no one else had already laid claim to the same area and that it contained a valuable resource, perhaps a precious mineral. Staking your claim later became a figurative idiom for asserting your right to something.

The claims we have been talking about in this book are obviously of a different sort, but hopefully the analogy is clear. If you find something valuable and want to stake a claim to it, you must prove that no one else discovered it first. That is done in the Introduction. Then in the second division you mark the boundaries of the claim, separating it from the work surrounding it. Finally, you must verify the value of what you have discovered. If an idea is not valuable, why should it be published, even if it is different? Thus the last two divisions of an article not only report and analyze results, but more importantly seek to convince readers that the contribution is valuable.

In this chapter we consider the last two divisions together for several reasons. Most importantly, as explained in Chapter 3, neither Results nor Discussion by itself constitutes a complete argument. Moreover, Results and Discussion are combined in many science and most engineering articles. Finally, although the same components tend to be present, the assignment of components to divisions is different. Although IMRaDC is like IPTC in overall structure, the Testing methods are usually in the Methods division, as in IMRD.

> We will consider the final two divisions together, as together they make and support a single claim, the contribution of the research.

6.2 Claim 3: the contribution of your work

Whatever the format of the last two divisions, their purpose is to show that the work makes an important contribution to the field. Recall the claim made in the last two divisions:

3 You have a good explanation for your data (IMRD)

or

3 Your solution is better than other existing solutions (IPTC).

Data patterns and interpretations are essential in IMRD, while comparisons are usually more important in engineering.

To support this claim, five components may be necessary: testing and analysis methods (6), data patterns (7), comparisons (8), interpretations (9), and conclusion (10). As with other divisions, there is a difference in emphasis between IMRD and IPTC.

In an IMRD paper two of these are always required: a summary of the data patterns (7) in the Results and interpretations of those data (9) in the Discussion. Comparisons (8), often in the form of statistical tests, may be included in either or both, while the last paragraph of the Discussion usually highlights the conclusion (10). Testing methods (6) are usually found in the Methods division.

In IPTC, testing methods (6) are usually found in the same section as the results of the test. Comparison of the present design with the previous best design (8) is normally required to prove that the new solution is better in some way, and some visualization of the data patterns (7) is needed to conduct that comparison, followed by a conclusion (10) to tie the paper together. Interpretation of the data (9) may or may not be required.

In both science and engineering these divisions frequently use an extended style, with initial introductory paragraphs for at least some sections. An introductory paragraph may look forward, summarizing that section, providing a general overview before specifics. On the other hand, it is also quite common to begin the third division with an introductory paragraph that looks back to summarize the research aim and methods. After reading the abstract, experienced researchers may jump straight to the results, since they are already familiar with both the literature and methods in the field and want to see how the results compare with established methods. A one- or two-paragraph summary of the research aim and methods is sufficient for them to understand the goal and approach. On the contrary, novice readers may have gotten so bogged down reading the details that they will have lost sight of the overall goal. Including a short summary thus accommodates their needs as well.

In these divisions it is common to have more than one component per paragraph. List all that you find.

Data from the present work, often presented in tables or figures, become far more important in these divisions as support for the component claims. Nevertheless, past research will need to be mentioned if there is any comparison with previous work.

In doing your component analysis of this division you may find *more than one component in a paragraph.* In IMRD a data summary may include statistical comparison, or comparisons may be interspersed with interpretations. In IPTC it is common to find a description of the test, a sentence pointing to a graph or figure depicting the results, a summary of those results, and comparison with previous studies, all in a single paragraph.

6.3 Component 6: testing methods

As discussed in the previous chapter, the types of tests typically conducted in science and engineering differ. Testing methods are usually described in the Methods division in IMRD, often as a separate subsection, but in IPTC this component is commonly found in the Testing division.

Frequently in IPTC several tests will be performed, with the testing method, data patterns, comparisons, and interpretations for the first test reported in one section before the same sequence is followed for another test in the following section. Sometimes all four components are described in only one or two paragraphs.[1] For example, the following sentences, from two successive paragraphs of an engineering article (Zhou & De la Torre, 2016), would be identified as components 6, 7/8, 8, and 9, respectively.

- We qualitatively evaluated our method with a leave-out scheme . . .
- Figs. 7a, 7b, 7c, and 7d qualitatively compare our methods with . . .
- Our method consistently outperforms . . .
- This is not surprising because . . .

Thus a good starting place for your component analysis is to identify what tests are performed and where the data for each are reported. Then consider whether the other components are in the same section as the data or a separate location. This may be easier than trying to identify the components by reading through the paper in order. Remember too that sometimes the Process is reported as a series of steps with intermediate testing before deciding on the next step. In this case there may be little or no Testing at the end, so the two divisions may be combined.

> **Begin your component analysis by identifying the tests performed and where the data for each are reported. Then look for related comparisons or explanations.**

6.4 Component 7: data patterns

Some sort of data summary is required in any research article, unless the research is purely non-empirical (see section 2.4).[2] Usually not all the data collected are reported, only those important to the argument, and results are often reported as summary statistics (e.g., mean or median) rather than raw data.

In naming this component, I have purposely chosen to use the term pattern, because data are rarely useful as evidence until a pattern is discerned. I have used the plural because usually more than one pattern is reported.

> Patterns must usually be identified for data to count as evidence. Figures and tables make those patterns easier to visualize.

A pattern may be shown in many forms, often including tables and figures since these make it easier to visualize patterns. Collectively, tables and figures are called graphics.[3] Thus, any paragraph in these divisions referring to a graphic should be identified as Component 7, whereas figures in previous divisions may show an overview of the process, or perhaps framework.

Data are usually measured in some way, using one or more of the testing methods (Component 6). As shown in Figure 6.1, the results may include not only data patterns (7a), but manipulation of that data after collection to make it easier to visualize the pattern (7b). Exceptions to the pattern (7c) may be even more important than the pattern itself, giving clues about how to interpret it. This could include possible sources of error, for example possible human or machine errors during the collection process, or outliers that do not fit the expected pattern and may lead to new insights. Sometimes the data may be placed into groups, with a different pattern in each group (7d).

> One paragraph may include more than one component, especially in IPTC, and should be marked accordingly.

In the engineering literature it is common for the data pattern (7) to be presented primarily, if not exclusively, in graphics. In science it is usually necessary to summarize these tables or figures in the text, or point out some important feature, but engineering articles sometimes point to a graphic without further comment. A graphic may also contain a comparison (8).

7) Data patterns

a. General pattern
   Claim: A pattern can be discerned in the data
   Support: Graphical representation or data summary
b. Data transformation
   Claim: The pattern is easier to see after appropriate procedures
   Support: Description of data handling or transformation methods
c. Exceptions to pattern
   Claim: Some data do not fit the overall pattern
   Support: Identification of possible exceptions, outliers, or data errors
d. Classification of patterns
   Claim: The patterns can be summarized into different categories
   Support: Similarities in patterns

Figure 6.1 Common sub-components of Component 7: data patterns.

Sometimes this is an explicit comparison between different methods described in the current article or comparison with results from previous studies. There may also be implicit comparison with a known goal in the field, for example how closely the results adhere to an established standard or the degree of similarity between expected and measured values. The text may also include an explanation of the reason for any differences (9). Thus, it is quite possible for there to be several equally prominent components in a single paragraph, which is far less common in science. Each component present should be marked.

In addition to mention of tables and figures, the following component markers indicate data summaries. The first two describe something about the data directly (7a), whereas the third describes a transformation (7b), and the fourth potential problems with the data (7c). The last, classification of the data patterns into categories (7d), is actually an inference that could be placed with Component 9, but is often reported in the results.

- We observe that . . .
- In all cases the proposed method correctly detects . . .
- Values were standardized by . . .
- Possible sources of error in the measurements include . . .
- Four different "profiles" emerged from the data . . .

6.5 Component 8: comparisons

Comparisons involve inferences, and should in principle be in the Discussion, but often are found in graphics with the current Results.

Comparisons pose a problem for traditional IMRD structure. In principle, evidence should be presented in the Results but inferences, including comparisons, in the Discussion. In practice this is nearly impossible, since data are normally presented only if a pattern is identified, but pattern identification involves inference. Furthermore, statistical tests rely on inferences, but test statistics are often shown in the same table as the data. A graphic may present not only the current test results, but also results from previous studies for comparison.

It might seem that there would be a standard format and location for reporting such an important category, but since articles are structured as arguments, good authors will present things in the order that makes the strongest argument for their case rather than following a set of rules for where things ought to go. Thus, instead of prescribing where comparisons ought to be placed, it makes more sense to find where they are in your articles and why.

Different types of comparisons are shown in Figure 6.2. Often there is a comparison of different treatments, designs, or parameters investigated in the current article (8a). Results may be shown in a table or graph, with or without statistical testing, as in the following:

- Figures 14 and 15 schematically depict the proposed one-hour-ahead and one-day-ahead forecasting model, respectively.
- Comparing the results between years, . . .
- Fig. 7(c) verifies the reduction of training error . . .

8) Comparisons

a. Within the article
Claim: One treatment or design is better than others
Support: Graphics, data summary

b. With prediction based on hypothesis (IMRD)
Claim: Data [support/do not support] hypothesis
Support: Statistical testing

c. With expectations
Claim: Data [match/differ from/exceed ...] expectations
Support: Graphics, data summary

d. With theoretical ideal (IPTC)
Claim: Results can be compared with a theoretical (mathematical) ideal
Support: Basis for ideal and comparison with it

e. With previous work
Claim: Data [support/refute/improve on ...] previous work
Support: Past research compared with present, in graphics or text

Figure 6.2 Common sub-components of Component 8: comparisons.

Comparison with a hypothesis-based prediction and its accompanying use of statistical testing (8b) is far more common in science than engineering. Studies that use statistics should report the hypothesis and prediction, the experimental design (assignment of treatments to subjects, repetitions, etc.), the test used and degrees of freedom, the results, and the significance. If your exemplar article uses statistics, read section 8.3 and try to identify where these various elements are in the article. Unfortunately, sometimes important elements are left out.

The location of these elements varies greatly. A prediction may be stated as part of the experimental question in the Introduction, with the experimental design in the Methods, or immediately before the data in the Results. It is also possible to give the results first, leaving the prediction and statistical tests until the Discussion. Statistical comparisons are often closely linked with the Results, and thus reported there, even though the statistics are by nature inferential, and therefore should "properly" be placed in the Discussion. A single graphic often reports both data and a statistical comparison, either as error bars on a graph or standard deviation or significance tests in a table.

The following component markers indicate statistical comparison:

- The mean + standard mean error (SEM) . . .
- The 95% confidence intervals of the mean . . .
- . . . was significantly higher than . . . ($p < 0.001$)
- The *p*-value for the entire data set was calculated as . . .

Note that not every occurrence of the words "statistics" or "significant" indicates statistical testing, especially in engineering, where these words are used much more loosely than in science. For example, the following sentences have nothing to do with statistical testing. "Substantial" or "considerable" would be a better word choice in the first, since no statistical test of significance was conducted. In the second, the intended meaning is "values." In the last, the meaning is "important."

- This is a significant improvement over previous methods.
- The simulated statistics match the measured statistics very well.
- It is significant to note that . . .

In many fields of engineering there is no hypothesis and therefore no prediction, and no repeated measurement or statistical testing.[4] Nevertheless, it is very common to compare the results obtained with the results expected based on a model (8c).[5] The comparison may be between measured results and a simulation or between an observed pattern and the expected pattern. Often this is shown in a table or figure rather than in the text. The following component markers all indicate a comparison with the model or expected values:

- The . . . distribution agrees well with simulations.
- The maximum measured is . . . , whereas the simulated maximum is . . .
- The calculated value from the simulation will be compared to the measured value . . .
- The estimated and actual values are very close . . .
- The average error is less than . . .
- As expected, case 1 has the largest . . .
- . . . are compared with the performance of the reference geometry.

Again, single words cannot be used to assign a component. The following example mentions "expected," but in this case it is talking about a theoretical expectation used to determine what value to give to a mathematical term, which has nothing to do with data or comparison, and would be classified as 5d.

- As mentioned before, exponent γ in these equations is expected to have a small value to smoothen the effect of the amount of length or resource constraint violations. Therefore, we have set $\gamma = 0.1$ in this paper.

Comparison may also be made with some optimal standard, a mathematical or theoretical ideal (8d). This is more common in engineering but may occur in sciences like physics when an optimum can be identified.

- Our technique can generate mappings that are close to the ideal performance defined earlier.
- We tested against optimal parameters without noise.

Whereas comparison with previous work (8e) is optional in IMRD, it is nearly obligatory in IPTC, for the obvious reason that the author claims the new design is in some way better than previous designs. Sometimes testing is done using established

Statistical comparisons are more common in science; comparison with expectations or an ideal are more common in engineering. Comparison with previous research is found in both.

methods in the field. Other times it involves coming up with a new test that better demonstrates the advantage of the new design. The following component markers identify comparison with other research. The first four express the intent to compare; the next four that the proposed solution is better than others. The final two present confirmation of previous research, which is more common in IMRD and likely to be reported in the Discussion.

- Our method is compared with the method proposed by . . .
- We compared our algorithm to four others published in the literature.
- Our methods were compared with three baselines, . . .
- Since previous methods reported their accuracy according to two different error metrics, we show our results in both ways for comparison purposes.
- Our method consistently outperformed . . .
- As expected, the common-centroid error obtained by the proposed method is smaller than the one obtained in [11].
- The best method, however, turns out to be our transformation . . .
- However, for [situation], this scheme would outperform the control.
- This confirms the results of previous study . . .
- Results show . . ., which agrees with the results in . . .

6.6 Component 9: interpretations

Items typically found in the Discussion are listed in Figure 6.3, all of which are far more common in science than engineering. In scientific articles, discussion of the cause of the results is mandatory. This is related to hypothesis testing and comparison of the actual results with the predicted results, but often goes further in discussing whether there could be other possible explanations of the same results, or how certain you are of your explanation. Depending on the study, it will frequently include limitations on how broadly the results can be applied.

Discussion or explanation of the results is mandatory in science, but often more limited in engineering.

In engineering the argument is more likely to focus on the limitations—that the new design is better in some area, just as good in some other, and perhaps not as good in yet another. It will include situations in which the new design would be expected to have an advantage and why. If practical applications have been tested, these may also be discussed. It is less common to

9) Interpretations

a. Analysis methods
Claim: Interpretation of the data was carried out in a reasonable way
Support: Details of analysis method, assumptions
b. Cause of results
Claim: There is a reasonable cause for the data
Support: Explanation of data pattern based on model or design goal
c. Elimination of causes (IMRD)
Claim: There is no better cause for the data
Support: Lack of support for alternative hypotheses or explanations
d. Practical applications (IMRD)
Claim: Understanding gained through this research has practical use
Support: Potential applications of the new knowledge
e. Limitations
Claim: Conclusions are valid within a limited range of conditions
Support: Limited range of testing or advantage

Figure 6.3 Common sub-components of Component 9: interpretations.

speculate on the underlying cause for the improvement, since often it was expected based on theory.

Sometimes an author must justify the methods used for analysis (9a). This may be based on standards, methods used by previous researchers, or argument that a certain method yields better results. When present, this is frequently found where the test is described.

- In order to assess . . ., it is necessary to perform . . .
- Therefore, it was necessary to compare . . . when evaluating
- Better resolution was typically obtained . . .
- We opted to use . . ., as this method allows . . .

Explanation of the cause (9b) is not a direct observation, so it must be stated as the most likely interpretation (see section 22.3.1), as in the following:

- For this reason, . . .
- The reason our approach can . . .
- This is on account of . . .
- . . . is achievable because . . .
- . . . is mainly caused by . . .
- This phenomenon is attributed to . . .
- A reasonable explanation of the increase . . .
- . . . appears to have
- It appeared that . . . due to . . .
- It appears from these initial results that . . .

- . . . which can be interpreted as . . .
- . . . could be explained by . . .
- It is plausible that . . .

Sometimes it is not enough to speculate on a cause. Other possible causes must be eliminated (9c). Even in science, this is common only in cases where another possible explanation seems likely.

- Therefore, we cannot attribute the increase . . .
- . . . does not seem to be relevant.
- . . . appears to have limited capabilities . . .
- On the other hand, it is possible that these effects were nevertheless responsible . . .
- The alterations described in the present study do not represent . . . and can be interpreted as . . .

In science it is also common to consider possible practical applications of the present research (9d).

- One of the most important applications . . .
- The approach can be applied to . . .

In the following component markers, the focus is on feasibility or limitations of the design (9e).

- Limitations of the current study . . . include small sample size . . .
- However, there are methodological concerns suggesting that . . . may be underestimated.
- However, it is very difficult to . . .
- The estimated . . . are inaccurate because . . .
- There is a small discrepancy . . . , which may be due to . . .
- The main reasons for this discrepancy are assumed to be . . .
- . . . is the main drawback of the proposed solution.
- . . . the proposed algorithm is sensitive to initialization.
- Our . . . algorithm can be confused when the assumptions made in this work are violated.

6.7 Component 10: conclusion

> The Conclusion verifies that the research goal was achieved.

The final component, shown in Figure 6.4, will comprise the Conclusion section, if there is one, or the last paragraphs of the Discussion. It summarizes how the article achieves the research goal and therefore makes a valuable contribution to the literature of the field. If the research goal had several questions or sub-points, each should be addressed separately.

In both engineering and science the wording of the conclusion may be nearly identical to the research goal. While the goal states, "this is our aim," the conclusion merely changes

10) Conclusion

a. Summary
Claim: The research goal has been achieved
Support: Summary of research, answers to questions or new design

b. Achievement or advantage (more common in IPTC)
Claim: Solution improves on past work
Support: Contribution of research

c. Future work (more common in IMRD)
Claim: Remaining questions can be answered or further improvements made
Support: Planned or possible future work

Figure 6.4 Common sub-components of Component 10: conclusion.

it to "we achieved our aim." Far greater difference is found in the social sciences, which tend to word the research goal as a question, with the answer presented in the conclusion.

A summary of the article, demonstrating accomplishment of the research goal (10a), is common in both IMRD and IPTC format. Perhaps because of the competitiveness and rapid change in engineering, the second sub-component is very common, highlighting the immediate advantage of the new design (10b). IMRD articles frequently include future work (10c), which goes beyond the immediate results of the paper to suggest related research planned by the authors or that others could contribute to. In many cases the conclusion will include two or even all three of the sub-components.

The following component markers introduce summary statements (10a). They point to accomplishment of the research goal, without explicit comparison with other research. This is common in both IMRD and IPTC.

- Our results show . . .
- In summary, we have shown that the properties . . .
- We have presented results of . . . experiments . . .
- We estimated the thicknesses . . .
- A novel system for . . . was presented. A novel strategy for . . . was proposed.
- A novel dual-mode control scheme . . . has been developed.
- The novelty of this work is the design and fabrication of a complete system for measurement of . . .
- A new algorithm has been proposed . . .
- We develop a robust, optionally fully automatic method that utilizes . . .
- The comparisons above indicate the feasibility and accuracy of the combined method.

Achievement or advantage (10b), more common in engineering, emphasizes why the current design is better than previous solutions. The following component markers highlight the advantage:

- Some specific features and contributions of the paper include: [list]
- Through extensive experiments, we have demonstrated the effectiveness of our algorithm.

- We have demonstrated that . . . reduces energy cost and provides significant energy savings.
- We have shown how to determine the minimum number of elements . . .
- To overcome these limitations, in this paper a novel technique has been proposed . . .
- We showed how . . . outperforms state-of-the-art approaches to . . .
- Unlike existing approaches, . . . incurs no design modifications and comes at a lower test cost.
- The main advantages over previously established methods are as follows. [list]
- The proposed design methodology can also be applied to different types . . .
- Another potentially beneficial application . . . is the possible use for . . .

The following component markers talk about future work (10c). While this is more common in IMRD, it is also found in IPTC when the work is currently underway.

- Our study poses new important questions that need to be addressed.
- Research on this . . . can be extended further to reduce . . .
- The optimization of this design to achieve thinner modules with higher gain is currently under investigation.
- Future work entails making our method more efficient and supporting a greater variety . . .
- In the future, in order to validate the expected value of . . . in a practical setting, we plan to implement and evaluate . . . on a real storage system.

6.8 Comparison of IMRD and IPTC

As has been mentioned several times, the location of some components differs in the two formats. Although IMRaDC is structurally similar to IPTC, Testing methods still tend to be found in the Methods division. Most of the components are present to some degree in all research articles, but because of the difference in overall goal, engineering is more likely to focus on Components 7 and 8, while Components 7 and 9 tend to be more important in science. Particularly in IPTC, two or three components may be found in a single paragraph. Some sub-components are more common in science, others in engineering. Nevertheless, there is considerable variation between papers.

Since one article often contains multiple tests, each with their own results and discussion, this almost always leads to some sort of cycling through the components in the final two divisions. Sequencing strategies will be addressed in section 9.5.

Exercise 6.1 Results and Discussion or Testing and Conclusion division components

(1) Identify the components in each paragraph in the last two divisions of your first exemplar article and the component markers that help you identify them. Summarize this in a table.

Hint: Remember that there is frequently more than one component per paragraph in these divisions.

Hint: As you start, look at the section and subsection titles. Do they give you an idea of what you will find in each section?

Hint: It may be helpful to first identify what types of testing are performed, then where the results of each test are reported, and finally where the comparisons and interpretations related to that data are found, which may be in the same paragraph, the same section, or a different section. If you can identify the locations for the first test, other tests will probably follow a similar pattern.

(2) Does the order of components generally follow the order listed? Do any of them repeat cyclically? Can you figure out a reason for the order?

Hint: The order may repeat cyclically, e.g., $data_1$/$explanation_1$, $data_2$/$explanation_2$, either because it is easier or because the authors need the results of test 1 before they choose what to do for test 2. For further discussion of sequencing methods, see section 9.5.

(3) How close is the wording of the Conclusion to the research goal in the Introduction? How similar is the wording of the same idea in the Abstract?

Hint: Recall that this is the main claim of the paper.

(4) Is there an introductory paragraph at the beginning of any section? If so, does it point back to the research goal or forward to the contents of that section?

Teaching tip: Questions 3–4 are necessary for an advanced class ready to write their own paper, but could be eliminated for a basic class.

Evaluate your understanding

(1) Why are the last two divisions discussed together in this chapter?
(2) What is the purpose of each component found in the last two divisions?
(3) How do science and engineering articles differ in the last two divisions?

Homework

(1) Repeat Exercise 6.1 for your other two exemplar articles.
(2) Write general claims about the structure of your exemplars, with support and exceptions.
(3) Compile a complete description of the structure of a typical article in your research field, based on your analysis of the three exemplar articles.

Hint: Number any tables and figures in your description sequentially, following the title format in your exemplars, and make sure you refer to all the tables or figures in the text of your description, either summarizing them or pointing out certain information the reader should pay attention to. For example, you might say, "As shown in Table 4, . . ." or ". . . as can be seen in Table 5."

Teaching tip: After students turn in a complete description, it is useful to have them do a revision, based on your comments and/or peer review. They will be able to see the improvement, and will understand the process when working on their own writing later.

Supplemental material: Sample component analysis.

Notes

1 I have also seen this in social sciences such as economics and management, where a new method is presented in the equivalent of a Process division, followed by a section comparing it to previous methods.
2 In this case, the Results and Discussion will be totally absent, apart from justification, limitations, and other comments interspersed with the mathematical proof.
3 See section 8.4 and Chapter 24.
4 These are, however, common in biomedical engineering.
5 This is essentially identical to comparison to a prediction (8d), as both compare to a theoretical value, but for the sake of student understanding I have chosen to list it separately since the terms hypothesis and prediction are commonly associated with statistical testing and are therefore rare in engineering.

Reference

ZHOU, F. & DE LA TORRE, F. 2016. Spatio-temporal matching for human pose estimation in video. *IEEE Transactions on Pattern Analysis and Machine Intelligence*, 38, 1492–1504.

Chapter 7

Evidence from past research

7.1 Connecting with accepted evidence

In the previous three chapters we examined the general claims commonly made in research articles. Now we will look in more detail at the evidence and reasoning used to support those claims. Evidence can come from past research, the focus of this chapter, or present research, discussed in the next. Reasoning, including ways to help readers follow the argument, will be the subject of the third.

> **Past research may be cited throughout an article to support various claims, since publication implies acceptance by the research community.**

References to past research establish a connection not only to a research community, but also to the evidence presented in those papers. Publication implies a degree of acceptance by a community, so published research can be cited by other scholars as evidence in their own writing. Previously published work is less subject to scrutiny than new evidence from the current research, but cannot be used to support every claim.

Before going further, we need to distinguish two terms. A *citation* is a marker in the text that points to previous work. Author-date citation format, common in science and social science, places the name of the author and the date of publication in parentheses, while numerical format, common in engineering, uses a numeral in superscript or brackets. The citation points to a *reference* listed in the References, Works Cited, or Bibliography at the end of the article. In Chapter 25 we will examine reference and citation formats, but for now we will consider only their use in the text used as evidence.

7.2 Evidence from past research

Past research is never mentioned for its own sake. To have value it must support the arguments made in the present article. Past research is essential for supporting certain claims and may be useful for others.

As discussed in Chapter 4, past research is used in the Introduction to establish the importance of and need for the research which leads up to the research goal. In IMRD this is the main function and location of citations, so textbooks on academic writing often discuss past research only in this context, with at most a passing mention in the Discussion section. Nevertheless, citations are common in all divisions of IPTC format except the Conclusion. They often appear in the Process as the framework underlying the current research and in Testing to compare the current design with past research.

7.2.1 Categories of past research

Past research can be mentioned for several purposes, listed in Figure 7.1. The first three occur in the first two divisions of an article, while the last may occur throughout, and is the only one that always requires a citation. The first two rarely have citations, so they must be identified based on the content. A widely used framework may be mentioned without citation, unless it is closely associated with a single researcher. As discussed earlier, the framework or approach used is sometimes not stated, but can be inferred based on the papers cited.

Background knowledge (I a) is general information the field considers a well-established fact. It is usually stated in the present tense, since it is considered inherently true independent of time. It may include anything from the needs of society to the specifics of the field, and usually meets the following four criteria:

(1) Written in present tense;
(2) Without citation;
(3) Not about new research presented in that article;
(4) Known and accepted by experienced researchers in the field.

Since it is so field-specific, there are no predictable sentence patterns, but the following would all be considered background knowledge. All were found in the Introduction and most are related to importance. Common in the first paragraph of engineering articles, this category appears to be less common in science, which prefers citations even for apparently obvious statements.

- A number of mmWave bands <u>are currently being considered</u> for global 5G networks.
- Power flow on overhead transmission lines <u>is limited</u> for a number of reasons.
- <u>An additional challenge</u> present in the telehealth environment <u>is</u> the large quantity of data which requires analysis.
- An attacker <u>could launch</u> a Denial of Service (DoS) attack by generating a large number of anomalous packets with different header fields.
- The SEE characteristics of insulator materials <u>are difficult to study</u> because the charge release is a gradual process over time.

I. Past research
a. General background knowledge for those in field (often without citation)
b. Definitions, terminology, or notation (with or without citation)
c. Description of theoretical frameworks, models, or approaches
d. Citation of specific methods, results, or conclusions from past research

Figure 7.1 Common evidence from past research.

- The storage mechanism consists in electrostatic adsorption of electrolyte ions on the electrode surface, which means that the capacity increases with the electrode-specific surface area.
- During the activation step, the oxidation effect of the activating agents involves an increase in O content and a decrease in C content.
- The electrocardiogram (ECG) is an important tool in assessing cardiac health.
- Breast cancer is the most common form of cancer among women worldwide.

Definitions (I b) may have citations if they were proposed recently, but not if they have become common knowledge in a field. In computer science they often occur in a separate section at the beginning of the Process division with the name Preliminaries. In other papers definitions may appear as a list at the beginning of the Process division with one definition per paragraph and the words defined in italics. Mathematical representations or notation, which may appear as an unnumbered section entitled Nomenclature placed before the Introduction, also fall in this category. Unless first defined in the current article, definitions are part of the Framework component. The following would all be called definitions or notation:

- Here we define: . . .
- A *timing path* is defined as: . . .
- *Definition 1 (Connected Graph)*: A graph is represented as . . .
- We represent graph vertices as positions in the image.
- We denote binary vectors in bold lowercase characters.

Citations will help you identify many but not all references to past research.

Framework (I c) refers to information other than definitions related to the theoretical framework of the present research. It may occur in the Introduction in IMRD but more commonly in the Process division of IPTC. We will discuss this more in the next subsection.

Citations of specific information (I d) are the easiest to identify, because of the citation, but it is important to determine what purpose they serve in the argument. The connection with various components will be described in the next subsection.

7.2.2 Past research as evidence for various components

Past research is frequently cited in the Introduction as evidence for importance or need.

Past research is frequently used as evidence in connection with certain components, yet rarely with others. This makes sense when you consider the purpose of each component.

Past research is frequently cited in the Introduction in both IMRD and IPTC. Here it shows either the importance of the research or limitations of previous research and therefore the need for the current work.[1]

Importance (Component 1) establishes why anyone should care about the research, either to publish it or read it. Just like promoting a tourist destination, if many people have gone there, pointing to the numbers is sufficient. Thus in engineering it is common to have strings of citations, but these quickly become cumbersome in the author-date citation format of science, as shown in the third example.

- Since then, numerous query authentication solutions have been proposed . . . [8, 14, 17, 21, 22, 29, 32, 33].
- Previous works on circuit-level modeling and simulation . . . [8]–[21] . . .
- Rheological properties . . . have been studied extensively in the last three decades [Doi (1980a); Osaki and Kurata (1980); Osaki *et al.* (1981); Venerus *et al.* (1990); Isono *et al.* (1991); Venerus and Kahvand (1994)].

It is also common to describe at least a few studies individually. Sometimes by only reading the description of each study it is impossible to tell whether the author intends to use it to support the importance or need of the research, but the intent may be clarified by other statements in the paragraph. For example, in two successive paragraphs of a science article we find the following sentences (Wen & Hua, 2009: 781–782):

- Thus, understanding the entanglement dynamics and structure is of great importance as the transient entangled network of the fluid affects the subsequent material deformation and stress relaxation.
- Thus, a better understanding about the impact of the transient entanglement structure is essential if one wishes to more accurately describe the chain dynamics and associated rheological properties in fast flows.

While importance and need are closely related, the first paragraph and the citations in it stress why we care about this topic (it affects important physical properties), while the second builds the case that more work needs to be done (to better describe how those properties affect flow). The following paragraph, which states the research gap and research goal, contains no citations. This makes sense because the claim is that there is no research in that area, but the current study will fill that gap.

- It appears to us, however, that no earlier experimental studies have conducted . . .
- In the present study, . . .

> **Related Work and Background sections in IPTC frequently are more related to need, but Theoretical Framework and Preliminaries to framework.**

As we saw in Chapter 4, some IPTC articles have a section immediately following the Introduction which contains much of the past research cited in the article. This is often named Related Work, Background, or a similar general title, although sometimes a section or subsection serving the same purpose will have a descriptive title. It appears to be most common in rapidly changing fields such as computer science and information engineering, ensuring that readers are aware of the most recent literature. Like a Literature Review in social science, this mainly discusses past approaches and their limitations, but the framework of the present study may be indicated as well.[2]

Framework is one of the hardest components for novices to identify. On the one hand, it may not be clearly stated by the author; on the other, readers may be unaware of or not have names for alternative frameworks. In the same way, a child may recognize a structure as a house, but only later acquire names for different types of houses in their area.

In science the framework may be implicit based on the choice of methods and analysis, particularly in highly specialized journals. In engineering, where there are usually several possible approaches to a problem, it is often found at the beginning of the Process division as information necessary to understand the present work. Usually in this case it can be identified by the presence of multiple citations, rare in the Methods division of science articles. Sections entitled Theoretical Background or Preliminaries may include definitions or nomenclature and a general description of the framework, with or without citations.

Citations may also occur in engineering in connection with the fifth component, details of the current research, when the new design incorporates a previously reported element without modification. Fewer citations are associated with Component 6, testing methods, since widely used standardized methods usually do not require a citation. For the same reason, citations are rare with these components in science.

Past research is frequently cited in the Discussion, but rare in the Conclusion.

Citations may also be found in the Testing or Discussion divisions. Citations will be mandatory if there is any comparison with previous research (8d), for obvious reasons, but less frequent with other types of comparisons, thus again more common in engineering. Citations will be more common with interpretations (9) in science, either to eliminate other possible explanations, confirm previous studies, or generalize the conclusions.

On the other hand, citations will rarely be found in conjunction with components that focus on the current research. Thus we would expect to find few associated with the research goal (3), data patterns (7), or the conclusion (10), and indeed this is the case. If data from previous studies are presented in the same data table or figure as the present study, this will have a citation. If the present work is an extension of prior work by the same author, the previous work may be cited in the research goal. The relative frequency of citations of past research in conjunction with each component in IMRD and IPTC formats is summarized in Table 7.1.

Exercise 7.1 Past research and components

(1) In your first exemplar, is there a separate section for definitions, related work, or theoretical framework? Is the title general or specific?
(2) Is accepted knowledge ever mentioned without citation (background, definitions, framework), and if so, where?

Table 7.1 Relative frequency of past research as evidence for each component

Component		*IMRD*	*IPTC*
1	Importance	++++	++
2	Need	++	++++
3	Research goal	-	-
4	Framework	++	++
5	Research details	+	+
6	Testing methods	+	+
7	Data patterns	+	+
8	Comparisons	+	++
9	Interpretations	++	+
10	Conclusion	-	-

(3) Where are the citations in your exemplars? What components do they seem to support?

Hint: As with earlier chapters, your exemplars may differ from the common pattern, and this is important to note.

Hint: You may find it useful to create a table showing how many citations are associated with each component.

7.3 Author-prominent vs. information-prominent citations

As we saw above, in every field some information is so well known and widely accepted that citations of specific past research are not needed. In your own writing you may wonder whether you need to include citations or not. The expectations vary somewhat between science and engineering; as always, your exemplar articles are your best guide to your field. If your exemplar articles mention the importance to society, definitions, or the name of some technique without citations, you probably can too. If other papers give a citation for certain information, you should too.

When citing an article, an author needs to decide which is more important in the situation: the author or the information. Author-prominent and information-prominent citations[3] differ in how much attention is given to the author of the article cited.

> In an author-prominent citation, the name of the author appears in the sentence, but in an information-prominent citation it does not.

In an author-prominent citation, the name of the author is an integral part of the sentence, often the subject, as in the following examples from "Spatial query integrity" (Hu et al., 2013):

- Papadopoulos et al. [26] designed a solution for . . .
- On the other hand, Ku et al. [13] proposed . . .
- The idea . . . was first introduced by Hacigümüs et al. [9].

In an information-prominent citation, the name of the author is not part of the sentence, as in the following examples from the same work:

- The first mechanism for verifying query results in *multi-dimensional* databases was proposed in [3].
- However, since [13] is not a deterministic solution, attacks may escape the auditing process.
- Aggregate signatures are provably secure [1, 18, 19] . . .

A variation of the information-prominent citation seems to be more common in mechanical engineering than other fields.[4] It does not use the author name, but inserts the word "Reference," as in the following sentences:[5]

- Results for stiffness design considering large displacement theory have been presented in References [21–23], . . .
- In this paper, the problem formulation from Reference [14] is advocated . . .

> The impact of using an information-prominent citation is much higher with numerical citations than with author-year format.

Note that there is a far greater impact of using information-prominent citations in the numerical citation format common in engineering than there is in the author-year citation format common in science. Consider the following two examples:[6]

- Based on [3], Cheng and Tan designed a mechanism for authenticating *k*NN queries on multidimensional databases, ensuring that the result set is complete, authentic, and minimal [4, 5].
- We plotted the epicenter of the mainshock estimated by Shin *et al.* (2000) and that of the aftershock estimated by the Taiwan rapid earthquake information release system (Wu *et al.*, 1997).

In the first example, without looking at the Reference section there is no way to determine who the author of citation 3 is, while in the latter we still know who both authors are, even though the first citation in the sentence is author-prominent and the second information-prominent, appearing in parentheses rather than the sentence itself.

> Author-prominent citations are rare in engineering, used mostly for the first paper in a field or the model on which the current research is based.

The relative frequency of author-prominent and information-prominent citations varies by field, but information-prominent citations predominate in engineering. Author-prominent tend to be used only for the first paper in a field or the framework on which the current paper is based.

In addition to citation type, the range of dates cited also differs. In the sciences there may be references to classic papers, sometimes 50 years old, but in engineering most of the papers cited are within 10 years, and it is rare to cite anything over 20 years old.

Exercise 7.2 Citation types

(1) For each section of your exemplar article, count the number of author-prominent and information-prominent citations. If the frequency differs, can you determine any factor that seems to influence the type of citation used?

Hint: Count by section rather than division, because the frequency may be far different in, for example, the Introduction and Related Work sections, both in the Introduction division.

(2) Do any information-prominent citations insert the word "Reference" or "Ref."?

Hint: Insertion of "Reference" is relatively unusual. It is only important to mention this in your description of your exemplars if it does occur.[7]

(3) How many references are less than 5 years old? 5–10 years? 10–20 years? Older?

Hint: Count the years between the reference and the date of publication of the article, not the date you are reading it. For example, a citation of a 2011 paper in an article published in 2015 would be four years old, no matter when you read it.

Hint: Again, consider the best way to present the information: table, text, or a combination.

7.4 Citation verbs

Word choice is important in any writing, not only to convey concepts accurately, but also to identify with a particular group and their shared view of reality. A change in government may be referred to as liberation or a rebellion. One side's freedom fighter is the other side's insurgent. In science the consequences of using the wrong term may not be as drastic as they could be in the political realm, but words are still important as a way of indicating your connection with an academic community.

What I will call *citation verbs report what a previous author did*. The preferred words differ by field, as illustrated in Table 7.2.[8] Nine of the words appear only once, comprising almost half of the top five words in these four fields. Also note that many common verbs used to talk about writing, such as write, state, and say, do not appear at all.

Citation verbs, used to describe what a previous author did, differ from journal to journal and carry positive, neutral, or negative connotations.

All words carry connotations. Consider the difference between the citation verbs in the following three statements:

- Jones (2012) discovered that A causes B.
- Jones (2012) stated that A causes B.
- Jones (2012) claimed that A causes B.

Table 7.2 Citation verb frequency in different fields of science and engineering

Rank	*Biology*	*Physics*	*Electrical engineering*	*Mechanical engineering*
1	Describe	Develop	Propose	Describe
2	Find	Report	Use	Show
3	Report	Study	Describe	Report
4	Show	Find	Show	Discuss
5	Suggest	Expand	Publish	Give

Selected and redrawn from Hyland (1999), used by permission, with additional data from Swales and Feak (2012: 213).

The first implies acceptance of the causality, while the last implies that the current author does not think the causality had been proven. The middle statement is neutral.[9] Citation verbs tend to have positive, neutral, or negative connotations, as shown in Table 7.3.

Note that far more of the citation verbs are positive or neutral than negative, since in engineering if an author disagrees with a previous work, they are more likely to simply not mention it than cite it negatively, unless the purpose of the article is specifically to challenge that previous method.

In engineering articles, most citations occur in the first quarter of the paper, in the Introduction and beginning of the Process division, but not all citations employ a citation verb. Author-prominent citations always contain a citation verb, either active or passive, as in the examples shown above.

> Author-prominent citations always contain a citation verb, either active or passive. Active voice further highlights the researcher as opposed to the research.

In author-prominent citations the tense and voice used also make a difference. Consider the difference in connotation of the following statements.

Table 7.3 Connotations of citation verbs common in electrical engineering*

Positive	*Neutral*	*Negative*
Demonstrate	Calculate	Assert
Determine	Conclude	Claim
Discover	Describe	Suggest
Establish	Develop	
Prove	Explain	
Resolve	Find	
Show	Present	
Solve	Propose	
Validate	Report	
Verify	State	

*Based on unpublished data of 30 articles from six journals.

Active voice:

- Wang [1] reports that . . . (present)
- Wang [1] reported that . . . (past)
- Wang [1] has reported that . . . (present perfect)
- Wang [1] had reported that . . . (past perfect)

Passive voice:

- It is reported by Wang that [1] . . . (present)
- It was reported by Wang that [1] . . . (past)
- It has been reported by Wang that [1] . . . (present perfect)
- It had been reported by Wang that [1] . . . (past perfect)

The active voice directs attention to the researcher, whereas the passive places more emphasis on the research. Therefore, the active voice is commonly used for landmark studies in a field to highlight the contribution of the author. Moreover, authors' names can be eliminated from passive constructions, but not active. Thus authors will sometimes refer to other studies in the active, but their own in the passive. In science and engineering it is common to find over half of the verbs in the passive.

> The tense used when citing previous studies also carries connotations about whether that research is considered valid.

> Information-prominent citations frequently do not use any citation verb, and are far more common than author-prominent in engineering writing.

Tense is also important, carrying a connotation about how researchers currently view that past research. Past tense is often used when citing a single study, because it describes the past action of one researcher at a point in time. On the other hand, the present perfect may be used when referring to several studies or a general area of inquiry because it describes what was reported over a wider time period. The present tense can be used to indicate something that is commonly held to be true by researchers in the field today. Conversely, the past perfect is used to report something that was previously believed, but no longer is.[10]

In information-prominent citations, passive citation verbs are more common, as shown above. However, unlike author-prominent citations, information-prominent do not require the use of a citation verb. In a typical engineering article, very few of the sentences citing previous work contain a citation verb.[11] This absence of a verb where there could be one is called a null occurrence. Null occurrences are very common when generalizing the work of many authors, in which case a string of citations may be listed at the end of a sentence or phrase, as in the first example below. Similarly, the second example below does not talk about what the author did in that paper; rather, the citation is a shorthand way of indicating a particular process, which is stated as an accepted truth. The third example below refers to the problem formulation from the reference without using a citation verb. The verb "advocated" refers to what is done in the current paper, not the reference cited, and is therefore not a citation verb.

- Aggregate signatures are provably secure [1, 18, 19] . . .
- However, since [13] is not a deterministic solution, attacks may escape the auditing process.
- In this paper, the problem formulation from Reference [14] is advocated . . .

Exercise 7.3 Citation verbs

(1) Look at each sentence in your exemplar article containing a citation. Underline all the citation verbs and make a list of the most common. Count the number of null occurrences, times where a citation is not accompanied by a citation verb.
(2) How many of the citation verbs carry a positive, neutral, or negative connotation?
(3) Look at the verbs you underlined. What tenses and voices are used in each section and for each verb? Is tense or voice used to mark anything the reader should pay special attention to?

Hint: How could you most effectively present the answers to these three questions: one table, more than one, a text summary?

Evaluate your understanding

(1) How does the location of past research differ in IMRD and IPTC, and why?
(2) Which components tend to use past research as evidence, and why?
(3) What differences are there between author-prominent and information-prominent citations, and when might an author choose to use each?
(4) How does the choice of citation verb, voice, or tense affect the connotation?

Homework

(1) Repeat the Exercises in this chapter for your other two exemplar articles.
(2) Write a brief summary of what you have learned about the location of past research and the use of citations in your exemplar articles.

Teaching tip: Sections 1–2 are important for all students. Sections 3–4 are more difficult and best left for advanced students.

Teaching tip: This chapter can be taught with the Introduction division, where past research and citations are most important. Sections 7.3 and 7.4 could also be taught in connection with section 22.3 (Connotation) or Chapter 25 (References).

Notes

1 Components 1–3 in this text correspond closely to the three moves of Swales' revised CARS model (Swales, 2004: 230) Swales considers past research an essential part of the first move and optional in the second. This does not seem to hold in IPTC, but further work is needed to document this.
2 Unpublished observations. As far as I know, no systematic research has yet been done on this.
3 Swales and others following him use the terms integral vs. non-integral citations (Swales, 1990: 148) I have chosen the terms author- and information-prominent (Weissberg & Buker, 1990), because the meaning seems more obvious to engineers.
4 This is called a hybrid form by Hyland (1999). Because the author's name is not mentioned and dropping "Reference" gives exactly the same form as information-prominent, I view it as a variant on that form.
5 From Pedersen, Buhl, and Sigmund (2001).
6 The first from Hu et al. (2013), the second from Satoh et al. (2001).
7 Linguists call the unusual situation "marked," which is marked or noted only when present.
8 Linguists call these reporting verbs (Hyland, 1999). Citation verbs are equivalent to reporting verbs, author focus, which are much more common than writer focus in science and engineering (Rau, unpublished data). The distinction between author focus and writer focus is not intuitive, particularly to those from an ESL background. I have chosen a term that designates the situation in which they are used.
9 It is also possible to insert an adverb before the citation verb to indicate the current writer's evaluation of the previous work, but this seems to be more common in the humanities and social science than engineering.
10 For more details and examples see Feak and Swales (2009: 52–54).
11 According to Hyland (1999: 249), roughly 40% of all citations in engineering used reporting verbs, but a recent preliminary analysis of electrical engineering articles showed that the frequency may currently be closer to 20% (Rau, unpublished data).

References

FEAK, C. B. & SWALES, J. M. 2009. *Telling a research story: Writing a literature review*, Ann Arbor, MI, University of Michigan Press.

HU, L., KU, W.-S., BAKIRAS, S. & SHAHABI, C. 2013. Spatial query integrity with Voronoi neighbors. *IEEE Transactions on Knowledge and Data Engineering*, 25, 863–876.

HYLAND, K. 1999. Academic attribution: Citation and the construction of disciplinary knowledge. *Applied Linguistics*, 20, 341–367.

PEDERSEN, C. B., BUHL, T. & SIGMUND, O. 2001. Topology synthesis of large-displacement compliant mechanisms. *International Journal for Numerical Methods in Engineering*, 50, 2683–2705.

SATOH, T., KAWASE, H., IWATA, T., HIGASHI, S., SATO, T., IRIKURA, K. & HUANG, H.-C. 2001. S-wave velocity structure of the Taichung basin, Taiwan, estimated from array and single-station records of microtremors. *Bulletin of the Seismological Society of America*, 91, 1267–1282.

SWALES, J. M. 1990. *Genre analysis: English in academic and research settings*, Cambridge, Cambridge University Press.

SWALES, J. M. 2004. *Research genres: Exploration and analysis*, Cambridge, Cambridge University Press.

SWALES, J. M. & FEAK, C. B. 2012. *Academic writing for graduate students: Essential tasks and skills (3rd edition)*, Ann Arbor, MI, University of Michigan Press.

WEISSBERG, R. & BUKER, S. 1990. *Writing up research*, Englewood Cliffs, NJ, Prentice Hall.

WEN, Y. H. & HUA, C. C. 2009. Chain stretch and relaxation in transient entangled solutions probed by double-step strain flows. *Journal of Rheology*, 53, 781–798.

Chapter 8

Evidence from current research

8.1 From past to present

Presenting known information before new information assists comprehension. This is a basic principle of communication. Thus citations of past research in the Introduction remind us what is already known by a research community, while new data from the current research are presented in the Results or Testing division. The transition between the two comes somewhere in the second division.

In science there is often a sharp transition at the beginning of the Methods division. Although the methods used may be similar to past work, the focus is the present research. In engineering the transition point may differ and the border may not be so clear. The position and clarity of the break seems to depend on at least two factors: (1) the overall structure of the article, and (2) how similar the new design is to previous solutions.

Sometimes in IPTC there is a sharp transition between past and present research at the beginning of a section, but not necessarily the beginning of the second division. Frequently the Process division begins with one or more sections or subsections on the framework (4) before discussing the present work (5) in a new section marked with words like "in our approach."

Other times there is more than one transition point. Often there are several new design elements. In this case the previous best design for each may be summarized before describing how it was modified in the present work, leading to a cyclical repeat of past and present work.

> In engineering there may be cycles between past and present research if the design includes several different elements.

Occasionally in IPTC the Introduction is very short, only one or two paragraphs, merely establishing the importance of the work and the research goal, with all the citations of past work in the Process division. Other times the Introduction may take several pages and include extensive detail not only on the connection between past and present work, but also a summary of the present design.

If the new design is very similar to earlier solutions, particularly in the case of extension of previous work by the same author, the transition between past and present may not be clearly marked, making it much harder to find.

Exercise 8.1 Past to present research

(1) Is there a clear break between past and present research? If so, at what point and what words mark the break, if any?

Hint: In science the break frequently occurs at the beginning of the Methods division. In engineering, look for markers such as "we propose," or "in this paper."

(2) Is there only one break point between past and present, or are there cycles?

Hint: This will help you find the break point(s) between framework and research details.

Evidence from the present research includes not only the data, but also details of how the research was conducted and the data collected.

8.2 Evidence from current research

When we think of scientific evidence, we normally focus on the data collected, but for the implicit claims of the second division (that the data were collected in a reliable way or the design is workable), the evidence is a detailed description of the procedures followed. Without this prior evidence, the data are worthless.

Figure 8.1 lists evidence related to the current study that support the claims of a research article. The first may occur anywhere throughout an article, especially in papers with mathematical formulas. The next four categories support claims in the Methods or

II. Present research

a. Numbered or bulleted examples or equations
b. Details and sources of materials
c. Details of present research design and procedures
d. Details of new design, intermediate testing of parts
e. Details of proof, lemmas, mathematical argument, algorithm
f. Detailed description of data collection and testing procedures
g. Data summary or highlight in text
h. Data patterns visualized in graphics
i. Data comparison, sometimes using statistical test results

Figure 8.1 Common evidence from present research.

Process division by giving a detailed description of the current research (5). The sixth supports the claims of the Testing component (6).[1] The final three, related to data pattern (7) and comparison (8), are found primarily in the last two divisions. Each will be described in the following sections.

8.3 Examples and equations

> Bullets or indenting make examples or equations easier to find, while numbers make them easier to reference.

Numbered examples are common in the social sciences but rare in natural science. In engineering an analogous feature is numbered equations. These can be referenced by number later in the paper or in future papers.

In other cases examples or equations are not numbered, but are formatted to stand out from the text, on a separate line or as bullet points. In social science the examples are frequently used as evidence in the Results or Discussion. In engineering, equations are more common in the Process, but may also appear as explanations in the Testing division.

Examples and equations are similar in format, if not purpose. Text is usually either indented and bulleted or centered and numbered sequentially along the right margin, in a format specified by the journal.

Exercise 8.2 Example and equation type and location

(1) Are there any numbered, bulleted, or otherwise highlighted examples or equations in your exemplar article? If so, in what division and in support of which component?

8.4 Evidence from research details

> Detailed descriptions identify Components 5 and 6, while the claims themselves are often implicit.

A description of the present research in the second division comprises Components 5 and 6. While the claim of this division is usually implicit, the detailed descriptions used as evidence are easy to find, so relatively little needs to be said about them.

In science, to prove that the data are reliable, an author must show that the materials were good quality, which is confirmed by listing the sources (II b). Even when following an accepted protocol, attention to detail (II c) shows that no errors were made in the procedure. Detail of the testing procedure used (II f) is one of the most important parts of the paper for reviewers in science.

In engineering, an author will undoubtedly purchase materials and use some established methods, but the details of the new design (II d) or mathematical procedure (II e) are far more important. As discussed in Chapter 6, the Testing method (II f) is likely to be found in the Testing division.

8.5 Data in text and graphics

> Graphics includes all the tables and figures in a document, whereas a graph is one type of figure.

> Tables and graphs are frequently found only in the Results in IMRD, but in IPTC illustrations are common in the Process division.

In the Results or Testing division, some data are summarized or highlighted in the text (II g), but graphics (II h) make quantitative data patterns easier to see. *Graphics* is a collective term for all the non-text information in an article, particularly tables and figures. It is *not the same as a graph*, which is one type of figure.[2]

There are differences from journal to journal in both the type of graphics used and their placement. Graphics attract attention and are one of the first things potential readers look at. Reviewers also evaluate the quality and usefulness of graphics when reviewing a manuscript.

Numerical tables are frequently used in both science and engineering to summarize data, while non-numerical tables, comparing categories or words, are rare. Graphs are the most frequent figures in scientific literature. Tables and graphs are most common in the Results or Testing, used to summarize data and show patterns or comparisons. Few are found in other divisions of IMRD.

In engineering other types of illustrations are also common, including schematic diagrams, photographs, computer-generated images, and pictorial models of abstract concepts such as waveforms. These are often found in the Process division, or even the Introduction, diagramming a new design, particularly if there are many distinct elements. For example, one of our sample engineering articles has two illustrations in the Introduction and five in the Process, in addition to 11 graphs in the Testing division (Hu et al., 2013).

Exercise 8.3 Graphics type and location

(1) Create a table showing the numbers of tables and figures in each division in the first exemplar article, separating them by type if useful.

Hint: Show where the graphics are first mentioned in the text, not where the graphic appears in the document.

Hint: You may find it useful to make separate categories for different types of tables and figures in your article.

Hint: What table arrangement would show this best? How will you incorporate information from your other two exemplar articles?

(2) Write a brief description, calling attention to any important features of the table, for example, "Over half of the graphics are in the Process division, with schematic diagrams being the most common type, as shown in Table x."

Hint: These descriptions will help you learn to summarize important features of data, an important skill in scientific and technical writing.

8.6 Data comparison and statistics

Like graphics, statistical tests (II i) are associated with the data, often in the Results but sometimes in the Discussion. They are far more common in science, explaining phenomena in the natural world, than in engineering, with its goal of solving a problem, often in a highly constrained environment where variation is assumed to be negligible. Nevertheless, some engineers use statistical tests in comparing the present to past methods, or when dealing with real-world applications.

While full coverage of statistics is beyond the scope of a writing textbook, their misuse is so common that a warning is required. If you use statistics, learn how to do them and write them up properly. Take a course, read a text, talk to experts, and understand why they are used, not merely how to plug numbers into a program. Even if you do not use them in your own research, if statistics are required in your research field you must know how to interpret articles you read. The number of statistical analyses in the literature that are either incorrect or poorly written is truly astounding. Please don't add to it.[3]

8.6.1 Types of statistics

There are two basic types of statistics: descriptive and inferential. Descriptive statistics are safe but of limited utility. They provide summaries of the current data, but no information on how those compare with other values. Statistical testing, on the other hand, refers to inferential statistics, in which the values measured are compared with other values or inferences made about a larger population based on the sample measured. They are tricky because there are many different tests based on the type of data and how they were obtained.

Descriptive statistics, common in both science and engineering, are a way of organizing and summarizing data, either numerically or graphically. The range, mean, median, variance, and standard deviation of a data set are descriptive statistics. Data can be displayed as bar graphs, histograms, box and whisker plots, or scatter plots. Correlation and least-squares regression lines are also descriptive, because they are mathematical descriptions of a data set, but cannot be compared with other data sets without further tests.

> **Statistical testing is common in science but rare in engineering. When conducted, it is imperative that details of the experimental design be included.**

Inferential statistics are based on hypothesis testing, using probability distributions to approximate random variation. Tests include confidence intervals, equality of means or the difference between means, goodness of fit, and analysis of variance. Inferential statistics are used in quantitative research to allow inference from the experimental results to a larger population or situation, and thus are far more common in science than engineering. They allow a researcher to estimate how much of the variation in experimental results is due to a real difference between treatments and how much is random variation due to environmental conditions or errors in measurement. Whenever statistical tests are performed it is vital to explain not only what tests were done and the results, but exactly how they were done.

8.6.2 Experimental design

How statistical tests were done is called experimental design, arguably the most important part of reporting statistics but the most often neglected by novices. It allows reviewers and readers to see if the methods you followed justify the claims you make about the results. It is important to report details such as:

1 how the test subjects were chosen and how treatments were assigned to them,
2 the experimental (independent) and response (dependent) variables and how each was manipulated or measured,
3 the number of times the test was repeated and degrees of freedom, and
4 the statistical package or procedure used to compare the results.

> The terms experiment, hypothesis, and prediction have both a common and technical meaning, which can lead to confusion.

Many statistical terms have a technical meaning which is slightly different from their use in common speech, leading to misuse in the science and engineering literature. Three of the most common related to experimental design are experiment, hypothesis, and prediction.[4]

An experiment, by modern scientific (not common) definition, involves changing the value of an experimental (independent) variable and measuring the effect on a response (dependent) variable, while holding all other variables constant. Usually it also involves repetition of each test to eliminate the effect of random variation in any real-world measurement. Many tests reported as "experiments," particularly in engineering, are not properly experiments. Misuse of the term perpetuates and exacerbates science illiteracy. In your own writing I encourage you to differentiate experiments from other types of testing, even if other authors do not.

Most experiments are designed to test a prediction, which is not the same as a hypothesis, a common misunderstanding that originates from the term hypothesis testing. Hypothesis testing involves making a prediction, a statement of what you expect to observe if a proposed hypothesis is true. Sometimes this is stated as an "if . . . then" expression: if my hypothesis (of the proposed explanation of the underlying mechanism) is correct, then my prediction (of what I will observe) will be correct.[5] Note that the "if" clause should be a statement of the hypothesis, not of what you do in the experiment. Let me illustrate by comparing the following examples:

- If I add a nitrogen-containing fertilizer, then a plant will grow faster or stronger;
- If nitrogen is a required element for plant growth, then adding nitrogen-containing fertilizer will help a plant grow faster or stronger.

The first statement makes a prediction of what will happen if I add fertilizer, but it is not a hypothesis because it does not explain why the plant will grow. The second statement properly pairs the hypothesis that nitrogen is a required element with the prediction that it will help the plant grow. The hypothesis in turn is based on an underlying model or understanding of how the factor being changed is related to the one being measured.

Exercise 8.4 Statistics type and location

(1) Are statistics used in your first exemplar article? If so, are they descriptive or inferential? In which division(s) do they occur?
(2) If they are inferential, do they provide all the necessary experimental design details?

Hint: Normally experimental design will be reported in the Methods in science, but is often absent or incomplete in engineering.

8.7 Pointing verbs

Since the data are often shown in graphics, the author must call the reader's attention to that information. Convention requires the author to refer to each graphic at least once in the text of the article, often using *pointing verbs.* Pointing verbs can *direct the readers' attention to a graphic* or *highlight certain information in a graphic.* These two functions are known as locating and highlighting, respectively.[6]

> There are various ways to point to graphics, with and without pointing verbs, but the present passive of show (as [is] shown) is quite common in engineering.

There are several ways of directing your readers' attention to graphics, depending in part on the style of writing. A topic sentence may point to the location of the data in a table or figure, with details the reader should observe in subsequent sentences. In many cases, locating and highlighting functions will both be found in the same sentence. The graphic may also be referred to in parentheses, without the use of a pointing verb. Examples are shown in Table 8.1.

Table 8.1 also illustrates another way sentences can be varied: by using active or passive voice. For the "locate only" category, the first example illustrates active voice. Properly, this sentence is incorrect, because a table cannot be an active agent and therefore cannot illustrate anything. Nevertheless, most journals accept this as a shorthand way of saying that the author illustrated the difference by composing the table. Some authors or grammatically inclined editors prefer to use the passive, as illustrated in the second example. This is a correct statement, with the agent, the author, unstated.

Table 8.1 Examples of different ways of referring to graphics

Category	*Examples*
Locate only	(1) Table 8.1 illustrates the difference between the three methods. In the simplest form only the location is given by the pointing verb. (2) Common pointing verbs are presented in Table 8.2. For most the author is the implied agent, the one doing the presenting.
Locate and highlight	(1) Location and highlighting can both occur in a single sentence, as shown in the second category in Table 8.1. (2) As can be seen from Table 8.2, the reader is the agent for a few pointing verbs, which also require a modal.
No pointing verb	(1) In summary, it is clear that there are several different ways of referring to graphics (Table 8.1). (2) By far the most commonly used pointing verb in electrical engineering is "show" (Table 8.2).

> The present tense is used when referring to graphics because although the work was done in the past, the graphic is visible to the reader in the present.

When location and highlighting occur in the same sentence the passive voice is often used, with the reference to the graphic placed in a subordinate clause, as shown in the second category in Table 8.1. Note that all pointing verbs are in the present tense, whether active or passive voice.[7] Even though both the gathering of data and production of the graphic occurred in the past, the data are currently being shown in the graphic, and can be seen by the reader in the present.

It is also possible to alter the sentence by changing the agent, as can be seen in the second example of location and highlighting in Table 8.1. In the first three examples, the implied agent is the author. By saying, "as can be observed," the action is shifted to the reader. This requires a whole different set of verbs, as shown in Table 8.2. It also requires use of a modal (usually "can"), since the reader must choose to look at what is illustrated.[8]

In addition to pointing to graphics, verbs can also *point to information in a different section*. Different verbs tend to be used for this purpose. Since they point to text rather than graphics, it is more common to use verbs such as present or describe, as shown in the last row of Table 8.2.

Exercise 8.5 Pointing verbs

(1) Look at each sentence in your exemplar article containing a reference to a graphic. Underline the pointing verbs and list the most common, separating those that point to tables from those used for figures.

Hint: There are many ways this could be presented. Is there any difference based on the type of graphic, or from division to division? Is number of occurrences important, or just rank order?

Table 8.2 Pointing verbs common in electrical engineering*

Category	*Pointing verbs*
Author agent (table)	Show Summarize List Give Present
Author agent (figure)	Show Illustrate Depict Present Plot
Reader agent (table or figure, with modal)	See Observe
Author agent (section)	Present Describe Discuss Show

*Rau (unpublished data).

(2) Count the number of null occurrences—times where a reference to a graphic is not accompanied by a pointing verb.

Hint: Look for references to graphics in parentheses.

(3) Are there any places where pointing verbs are used to point the reader to a different section of the paper? If so, what verbs are used?

Hint: You can find these by searching the article for "section."

Evaluate your understanding

(1) Why does science tend to switch suddenly from past to present evidence at the beginning of the Methods division, whereas the two may overlap extensively in the Process division in engineering?
(2) What is the purpose of highlighting examples or equations with numbers or bullets?
(3) How do science and engineering differ in the location and type of graphics, and why?
(4) If statistical tests are conducted, what information needs to be included in describing the experimental design?
(5) What is the difference between locating and highlighting graphics?
(6) Why is present tense always used with pointing verbs?

Homework

(1) Repeat the Exercises in this chapter for your other two exemplar articles.

Hint: Consider different ways of combining and presenting the information. What would make it easiest for a reader to understand the general pattern?

(2) Summarize what you have learned about the use of evidence from present research in your exemplar articles.

Hint: Compare your results with other students. You may be surprised how much difference there is from one journal to another, even within the same field.

Teaching tip: This chapter can be taught with either the second or third division, since this evidence is important in both. Related information can be found in Chapter 24 (graphics).

Notes

1 Evidence for the four sub-components of Component 5 are listed separately because each type of evidence is very different, while only one summarizes all the testing methods because no matter what type of testing is conducted, the procedures that must be described are similar.
2 The two letters separating graph and graphic make a vast difference. This is difficult for those working in English as a lingua franca, but there does not seem to be a better collective term.
3 Full disclosure: Although I am not a statistician, I have a personal stake in the proper use of statistics. The main contribution of my PhD research was proof that a statistical method used by 15 papers over 30 years underestimated gene number by at least an order of magnitude. Although I realized the problem, it took an expert statistician to determine the correct measure. Since then I have detected numerous statistical errors in other articles, either during review or after publication.
4 Two others, significant and correlation, will be discussed in section 23.3 in the context of word choice.
5 This has been called hypothetical inferencing. "If . . . then" clauses are used for different purposes in other fields, often with an "if (evidence) . . . then (claim)" structure, for example logical deduction in philosophy, legal consequence in law, and persuasive interpretation in English, with or without explicit statement of the reasoning (Nesi & Gardner, 2012: 118–119).
6 See Cargill and O'Connor (2013: 34).
7 Some students seeing a past participle ending in –ed or a modified form like "shown" will automatically identify it as past tense, but in the passive voice the tense is identified by the form of the verb "to be" associated with the past participle, so "is presented" is present and passive, whereas "was presented" is past and passive. In present passive with the conjunction "as," the verb "to be" may be omitted, "as (is) shown in . . .," making it even more confusing, since there is no apparent present tense in the construction.
8 This expression is properly called a passive modal simple.

References

CARGILL, M. & O'CONNOR, P. 2013. *Writing scientific research articles: Strategy and steps*, New York, John Wiley & Sons.

HU, L., KU, W.-S., BAKIRAS, S. & SHAHABI, C. 2013. Spatial query integrity with Voronoi neighbors. *IEEE Transactions on Knowledge and Data Engineering*, 25, 863–876.

NESI, H. & GARDNER, S. 2012. *Genres across the disciplines: Student writing in higher education*, Cambridge, Cambridge University Press.

Chapter 9

Clear reasoning

9.1 Follow the yellow brick road

Three hundred years ago the German philosopher Gottfried Leibniz wrote, "There are also two kinds of truths: those of reasoning and those of fact. Truths of reasoning are necessary and their opposite is impossible; truths of fact are contingent and their opposite is possible" (Leibniz, 1898: 235–236). Although the principle remains valid, the meaning of words has changed so much since the English translation of 1898 that it may require explanation. Truths of "reasoning" (those based on mathematical logic) are "necessary" (always true). Thus the Pythagorean theorem derived 2,500 years ago, like all other mathematical proofs, is still true today. On the other hand, with truths of "fact" (what we now call empirical truth), the "opposite" (a totally different interpretation) is possible. How that interpretation is done is the focus of this chapter.

Ideally, by following the reasoning in an article, all readers should be able to reach the same conclusion about the evidence that the author has. As an analogy, when Dorothy asks how to get to the Wizard of Oz, the Munchkins reply, "Follow the yellow brick road." If you stay on the road, you will reach your destination. The same should be true in an academic article. There should be no alternative paths (as Dorothy faced when she met the scarecrow), gaps to jump, or sharp turns that might be missed.

This chapter is by no means an exhaustive treatment of reasoning and organizational strategies, but will point out some features commonly found in science and engineering research articles. As shown in Figure 9.1, reasoning involves different types of logic and evaluation. There are also often signposts to help readers follow the argument.

9.2 Logical reasoning

> Deductive logic tests a hypothesis, inductive develops a new model, and abductive identifies the best of several explanations. All are common in science.

Three types of logical reasoning are commonly recognized in empirical research: deductive, inductive, and abductive. Most scientific research is deductive (III a), testing predictions made by a hypothesis or theory. Inductive research (III b) is the opposite, combining numerous lines of evidence into a new

Support: Reasoning

III. Logical
  a. Deductive: Testing predictions made by a general theory
  b. Inductive: Compiling observations to reach a general conclusion
  c. Abductive: Choosing the most likely of several possible explanations

IV. Evaluative
  a. Evaluation of previous research
  b. Evaluation of explanations
  c. Evaluation of current design

V. Organization
  a. Initial summary to reveal argument structure of subsequent section(s)
  b. Reminder of the research goal
  c. Concluding summary of argument in preceding section(s)
  d. Referring to arguments made in other sections

Figure 9.1 Common reasoning strategies.

model or theory. Abductive (III c) is sometimes called reasoning to the best explanation, choosing the most likely of several possible explanations for the data.

When developing the IPTC model presented in this book, I used inductive reasoning. I combined observations from many articles in different fields of engineering, trying to identify common components. In doing your component analysis, you are practicing deductive reasoning. Based on the overall structure of your exemplars, you stated a hypothesis that articles in your field follow either IMRD, IMRaDC, IPTC, or IPC format. Since then you have been testing that hypothesis, seeing whether the articles fit the expectations of that model or not. They may not, depending on both the research methodology and the individual author. As you try to assign components to each paragraph, you may engage in abductive reasoning. It may not be clear, for example, whether a paragraph at the beginning of the Process division should be called framework or research details, and you will have to choose based on the preponderance of evidence, even if one sentence does not seem to fit.

Generally logical reasoning is more important in science, with its focus on explanations of the data, than engineering. Each article usually can be identified as primarily one type of research, although a longer article or book could include more than one. Articles that perform hypothesis testing, most science articles, follow a deductive approach. Those that seek to develop a new model use an inductive approach. Observational studies that begin without a hypothesis may follow an abductive approach.

9.3 Evaluative reasoning

For both science and engineering, previous research is evaluated in the Introduction (IV a). In science the framework is often revealed by the author's attitude toward past research. Engineering identifies a need based on evaluation of the current best solution, noting limitations or weaknesses. Sometimes the order of presentation reveals the evaluation as well, with the method most similar to the current research, with the least limitations, mentioned last.

> Evaluation of designs in engineering often requires less discussion than evaluation of explanations in science.

In later divisions, whereas science uses data and statistics to test a hypothesis to determine whether an explanation is correct (IV b), engineering evaluates solutions in different situations to determine where each has a competitive advantage (IV c). Since there are objective measures of speed, weight, and other factors important in engineering, the evaluation often does not require the same type or degree of discussion or reasoning as science articles. The simple statement A<B may suffice in engineering, but science might require a statistically significant difference, a reasonable explanation for the cause of the difference, and elimination of other possible causes.

9.4 Extended vs. condensed style

To help readers follow the argument, authors often leave signposts to point the way. Two of the most common are introductory paragraphs and topic sentences. Using an analogy familiar to most readers, introductory paragraphs are like a GPS unit telling you, "In 100 meters, turn left," whereas topic sentences say, "Turn left." They occur at predictable points and help the reader navigate the text. On the other hand, a driver familiar with the route does not need the navigation, and may find it annoying and turn it off.

> Extended style is often used when introducing new material, but condensed for material familiar to most readers.

The two situations are analogous to extended and condensed styles in academic writing.[1] The former guides you through but can provide too much information if you know the territory. Thus extended style is common in sections introducing new material, while condensed is used for material familiar to most readers.

9.4.1 Characteristics of extended and condensed style

Differences between the two styles are summarized in Table 9.1. A plus sign (+) means an indicator is present, while a minus sign (–) means it is not present. The two easiest features to recognize are introductory paragraphs and topic sentences.

Although intermediate styles exist, many sections are relatively easily to categorize. The Methods division in IMRD and sections summarizing the framework in IPTC may be written in condensed style, since readers should be familiar with that material. Most other sections are likely to be written in extended style. In mathematics format, there may be a paragraph at the beginning of each section in extended style to explain why the procedure is needed, but the rest of the section will be in the highly condensed style typical of proofs.

Table 9.1 Variation in styles of scientific and engineering writing

Indicator	*Extended*	*Condensed*
Introductory or final summary paragraphs	+	–
Topic sentences	+	–
Details: definitions, examples, descriptions, process	+	–
Purpose, justification	+	–
Background knowledge assumed	–	+
Acronyms and abbreviations	–	+
Listing or citation without details	–	+

Adapted from Swales and Feak (2004: 227)[2]

> Extended style uses summary paragraphs, topic sentences, repetition, and explicit statement of purpose or justification, making it easier to skim and read.

In extended style a section may begin with an introductory paragraph, and each paragraph with a topic sentence. By reading only the first paragraph of each section and the first sentence of each following paragraph, a reader can quickly grasp the overall thrust of the article. Extended style tends to use short sentences and paragraphs, with one main point per paragraph and one important idea per sentence, another technique that makes it easier for the reader. It also adds non-essential words to make the argument clearer, such as examples or indicators of purpose like "in order to." Repetition of key words from one sentence to the next helps tie the thoughts together.

> Condensed style assumes a great deal of background knowledge, including familiarity with previous research and terminology.

Condensed style, on the other hand, lacks summary statements or repetition. The first sentence in a section or paragraph plunges directly into details. Ideas are presented linearly, using long sentences with many clauses and internal connectors (and, but), but fewer linking words or phrases between sentences (furthermore, despite). Several main points may be presented in each paragraph. It is impossible to skim for the overall meaning because each sentence addresses a different topic. Extensive background knowledge is assumed: both terminology and the content of previous papers.

9.4.2 Summary paragraphs

Summary paragraphs may be found at either the beginning or end of a section, and may serve at least four purposes, listed as the four subpoints of Organization (V) in Figure 9.1.

None are essential to the argument, but they ensure your writing is well organized and easy to follow.[3]

In many IPTC articles, the last paragraph of the Introduction section describes the organization of the rest of the paper (V a). This is also common in social science but rare in natural science. The paragraph begins with the following sentence pattern:

- This paper is organized as follows. In section II we . . .

It is also quite common to have a paragraph at the beginning of a section in the Process or Testing division that provides an overview of that section (V a), sometimes as a single paragraph before numbered subsections. This often includes the words "this section." It may also include words like "steps" or a series of sentences that indicate sequence, as in the last two examples below. A similar paragraph may be found at the beginning of a subsection. Examples of this sentence pattern include:

- In this section we present a brief introduction to this model.
- This section describes an overview of our proposed method.
- Our approach consists of three main steps.
- The two main components of our system . . .
- Section III-A introduces . . . Section III-B describes . . .
- In this section we present . . . We also . . . Finally . . .

As mentioned in Chapter 6, it is not uncommon in IPTC to restate the research goal at the beginning of the Process or Testing division (V b). This helps readers who may skip to a section of interest. The following examples restate the research goal:

- The goal of our work is to . . .
- Our research aims to provide . . .

Concluding summaries are common at the end of a section (V c) in the social sciences, but not in science and engineering. They rarely occur at the end of the Testing division to pull together different types of evidence.

It is also common in engineering to refer to something explained in a different section (V d), either earlier or later, as in the following examples:

- As previously stated in Section II, . . .
- We apply the . . . models proposed in Section III . . .
- We will discuss the impact of D in more detail in Section V.
- We discuss this issue further in Section 7.1.

Exercise 9.1 Identifying extended and condensed styles

(1) Does your first exemplar article use extended style, condensed style, or different styles in different sections?

Hint: Look primarily for summary paragraphs and topic sentences.

9.5 Sequencing strategies

Even within the expected structure of a research article, there are many ways of sequencing the same argument.[4] Which should come first, claim or evidence? Which evidence should be presented first? Present all the evidence before discussion, or alternate results and discussion? These are the questions that make writing a continuing challenge. Sometimes the right words come easily, but often even experienced writers will try several different orders before settling on one that seems to make the strongest argument. As Ernest Hemingway commented, "There is no rule on how to write. Sometimes it comes easily and perfectly; sometimes it's like drilling rock and then blasting it out with charges" (Hemingway, 2013).

There are several basic sequencing strategies, which can be combined either linearly or recursively (cyclically). Unlike components, evidence, or even the summaries discussed in the previous section, sequencing strategies are not identified by markers but by the order of components.

9.5.1 Basic strategies

At least four different sequencing strategies are commonly used in scientific and engineering writing, which I will call physical, temporal, relational, and mathematical. The first, *physical*, describes a topic in terms of its type, location, or parts. The second, *temporal*, traces a topic chronologically or by following the steps in a process, which are often but not always the same. For example, if two processes are happening at the same time, it would be possible to describe the order of events in both (chronological) or first describe the steps in process 1, then process 2. The third strategy, *relational*, involves organizing in terms of importance or some logical relationship. The fourth, *mathematical*, refers to the

Physical
- Type (equipment vs. consumables, hardware vs. software)
- Location (physical arrangement)
- Elements (parts in product)

Temporal
- Chronological (time sequence)
- Order (steps in process)

Relational
- Importance (priority)
- Comparison (evaluative)
- General to specific (deductive)
- Specific to general (inductive)

Mathematical
- Mathematical proof
- Algorithm description

Figure 9.2 Common sequencing strategies.

logical order of argumentation common to mathematics; for example, the order used for a mathematical proof or description of an algorithm. The various strategies and subtypes are illustrated in Figure 9.2.

Sometimes one ordering strategy is obviously best, but frequently several different orders could make sense. For example, in the Materials and Methods division, methods are a process, often calling for a temporal ordering, but materials could be listed by type, by order of use, or by importance. There may be a de facto standard strategy in your field that can be determined by analysis of your exemplars or the author may have freedom to choose.

9.5.2 Nested or cyclical strategies

Sequencing strategies are frequently nested, leading to cyclical repetition of common components.

Often two sequencing strategies are used in the same section. In this case, one is nested inside the other, leading to either a linear arrangement of the components or a cyclical repetition. The nesting strategy is often revealed by the section and subsection titles.

For example, in a traditional Materials and Methods section, Materials are described first, followed by Methods, then Testing methods.[5] If two types of tests are being performed, a second level of sequencing is required, but two nested strategies are possible, as illustrated in Figure 9.3, where the numbers indicate component numbers. The first leads to a linear sequence of the components, but in the second that sequence is repeated twice.

Nested alternatives can also be found in the Results and Discussion or Testing division. All the testing methods could be listed first, followed by all the results, then all the discussion (classic IMRD format). Alternatively, the testing method, results, and discussion of test 1 could be presented first, then the same for test 2 (common in IPTC format). The difference between the two is illustrated in Figure 9.4, where the numbers again indicate

5a Materials	Test #1
Test #1	5a Materials
Test #2	5b Methods
5b Methods	6 Testing
Test #1	Test #2
Test #2	5a Materials
6 Testing	5b Methods
Test #1	6 Testing
Test #2	

Figure 9.3 Nesting strategies for Materials and Methods in IMRD.

IMRD	IPTC	IMRaDC
6 Testing (M)	Test #1 (T)	6 Testing (M)
Test #1	6 Testing	Test #1
Test #2	7 Results	Test #2
7 Results (R)	8,9 Discussion	Test #1 (R&D)
Test #1	Test #2 (T)	7 Results
Test #2	6 Testing	8,9 Discussion
8,9 Discussion (D)	7 Results	Test #2 (R&D)
Test #1	8,9 Discussion	7 Results
Test #2		8,9 Discussion

Figure 9.4 Nesting strategies in IMRD, IPTC, and IMRaDC, by component and division.

component numbers and the letters in parenthesis the divisions. The second column is called a cyclical or recursive pattern, as the same components are repeated in a looping sequence, as opposed to the first column, where each component occurs linearly, although two different tests are discussed for each. A more complex pattern common in IMRaDC is shown in the third column. Like IMRD, the testing method is described in the Method division, but like IPTC there may be a cyclical repetition of components related to the Results and Discussion.

> Presenting topics in the same order in different divisions facilitates reading. Results often drive the organization in IRMD, but Process in IPTC.

The sequence chosen will depend on the format required by the journal, the topic and purpose of the author, and which will make the strongest or clearest argument. Generally, following the same sequence in the Process and Testing (or Methods, Results, and Discussion) makes it easier for the reader to follow. In IMRD the sequence will usually be determined by the order of presentation of data in the Results, while in IPTC the order in the Process is more important, as is fitting given the prominence of those respective divisions in each format.

For comparisons and interpretations it is common to use a relational strategy, because it is necessary to integrate various types of data. Since so many different orders are possible, this is one of the most challenging parts to write.

The Introduction is somewhat distinct from the rest of the paper. It seems to be more standardized across fields, with the importance of the work followed by the need for the work, then the research goal. The last three divisions focus on your research, and thus are quite variable based on both the field and the research methodology (see section 2.4).

Nevertheless, there are differences in the Introduction as well. The importance of the work frequently moves from general to specific, but needs to show the relationship between past and present research and therefore uses a relational strategy, with the approach closest to the present work often mentioned last. As we all know, relationships can be complicated and take many different forms, so there is far more variety in this organizational strategy. This is one of the reasons the Introduction is so difficult to write.

Exercise 9.2 Sequencing strategies

(1) What sequencing method seems to determine the order of the Methods or Process division in your first exemplar article? Is it linear or cyclical?

Hint: Look at the order of components you identified.

(2) Is this same sequence followed in the Results and Discussion or Testing?

Hint: Remember that more than one ordering scheme may be used, either nested (e.g., time sequence within each part) or different schemes for different purposes (e.g., part sequence when talking about the design, time sequence when talking about the test results).

(3) Can you speculate on why that order may have been chosen?

Hint: "No" is not an acceptable answer. Thinking about why it makes sense for this paper will help you decide whether the same method will make sense for your own writing.

Teaching tip: See class exercise 9.1.

Teaching tip: This section is more challenging for students than the basic analysis of components. It may be skipped without loss of continuity or covered in an advanced class only. It is also possible to simply identify linear and cyclical strategies without considering the reason the order was chosen.

Evaluate your understanding

(1) Describe the difference between deductive, inductive, and abductive reasoning in scientific research.
(2) How does evaluation differ in engineering and science?
(3) What are the main differences between extended and condensed style, and why would an author choose to use different styles in different sections?
(4) Why are there sometimes cycles of the same components?

Homework

(1) Repeat the exercises in this chapter for the other two exemplar articles.
(2) Summarize the reasoning and sequencing strategies common in your field.

Hint: Consider where to place this information in your overall description of exemplar articles. Should it be separate or combined with previous information? The order in which you discover something may not be the best way to present it to an audience.

Teaching tip: Use this as an opportunity to begin to talk about the rewriting process, organizing items by where they make the argument easiest to follow rather than following either a prescribed structure or the order in which the research was conducted.

Supplemental material: class exercise 9.1—comparison of sequencing strategies.

Notes

1 Terms from Swales and Feak (2004).
2 Swales and Feak also list "subsections" and "connectors or linking words" as evidence of extended style, but I have found these misleading for students. IMRD format papers often have subsections only in the Methods, with a very condensed style. The subsections contribute to that style by stating the content as a heading rather than a topic sentence. Similarly, connectors can take the place of repetition of words to shorten a passage.
3 Linguists call these metadiscourse, and give specific names to each type: future projections (V a), recapitulations (V b and V c), and location statements (V d).
4 Perhaps sequencing should be called rhetoric rather than reasoning, but that word has its own problems. One definition is positive—the art of effective or persuasive speaking or writing, but some consider it to have a negative connotation—persuasion with the intent to mislead or lacking evidence.
5 In science, Methods may refer to preparation of the study material, whereas Testing methods would be the measurements taken and statistical procedure employed.

References

HEMINGWAY, E. 2013. *The Cambridge edition of the letters of Ernest Hemingway*, Cambridge, Cambridge University Press.

LEIBNIZ, G. W. F. V. 1898. *The monadology and other philosophical writings*, Oxford, Clarendon Press.

SWALES, J. M. & FEAK, C. B. 2004. *Academic writing for graduate students: Essential tasks and skills (Vol. 1)*, Ann Arbor, MI, University of Michigan Press.

Part 3

Exploring different genres

Chapter 10

Argument structure in other types of writing

10.1 Variations on a theme

By now you should understand the argument structure of research articles in your field. At the beginning I told you it would be hard work but would pay rich dividends when it came to your own writing. Now you are ready to reap the reward.

There are many different genres of academic writing, but all are variations on a theme. Like research articles, all academic writing makes a claim or series of claims, sometimes explicit but often implicit, supported by evidence and reasoning. Perhaps you are ready to write a research article, a conference paper, or your thesis. Maybe you are an undergraduate, working on a project for a class. At some point you may write a grant application. All are arguments. Even a resume is an argument, not a list of past accomplishments.

> All academic writing is structured as an argument, so the principles learned from studying research articles can be applied to any writing.

In this chapter I will lay out some general principles for academic and technical writing. In the next five chapters we will consider how to adapt the same ideas to other types of writing, whether exemplars are available or not. Read whichever chapter is appropriate to what you need to write first, then skip ahead to Chapter 18 for steps in the writing process. You can return to other chapters in this part as you are ready to write other documents.

10.2 Planning to overcome difficulties

The same components found in research articles are found in many other types of academic writing, although not all to the same degree. Difficulties writing each component can be overcome with proper planning.

Three of the components—importance, need, and framework—are supported by evidence from past research. Extensive past research. Therein lies the difficulty. How many articles were cited in your exemplars? Thirty, fifty, one hundred? How many have you read? That is why it is important to quickly find the information you need from each article. From articles you have read you will be able to identify others you need to cite. You may not have to read the whole article if you know where to look for the information you need. On the other hand, make sure that you do not cite an article you have not looked at yourself, as citations are sometimes incorrect or misleading.

A second difficulty is deciding which past research to cite for each of those three components and how to order them. Summarizing an article in one sentence is a challenge. Stringing the sentences together is another. Even remembering what each article says becomes a challenge as the number increases.

> Adding information to your writing as you read helps you connect the article with known information, improving memory.

My suggestion is to begin with the general structure of the argument, as shown in section 19.2. Then as you read each article think which component or which other research it is associated with, write a one-sentence summary, and put it approximately where you think it will go. For example, write both the main advantage and limitation of a design, and place that sentence in a paragraph with others pointing to the need for your research. If it should go in two places, for example the limitations in the Introduction and comparison to your design in Testing, write two sentences. When you connect the information in one article with other information you know, it helps you remember it, something that does not happen if you take notes on each article separately.

After you have several sentences in approximately the right place, rewrite the paragraph to connect the ideas, reordering as necessary. You may not have written this way before, but it is effective when dealing with a large amount of information, especially in the age of computers that makes such revision easy.

The next component, research goal, is a far more important part of a paper than most novice writers realize. It determines what previous research must be included to show how your work adds to what is already known in your research community. It also determines what tests need to be conducted and how those are interpreted to achieve your goal. In short, it determines the contents of the whole paper. Remember that the goal needs to be stated in the format expected in your field. An author who writes research questions when the field prefers research objectives will seem like an outsider and risk rejection. By taking the simple action of transforming the questions into objectives, the rest of the paper may be identical but the chance of success far higher.

Introducing the framework presents different challenges in IMRD and IPTC. In engineering the challenge often is deciding how to distinguish past and present work. How much of the work you did is identical to what has been done before, and what is novel to your work? Both must be stated clearly. In science the framework is often implied by a writer's stance toward previous work, so care must be taken in the specific wording.[1]

For the research details and testing methods it is the details that make writing difficult. In science the language for writing these is so standardized that many phrases (but usually not whole sentences) can be copied from one paper to another without fear of plagiarism. Although this division is not the focus of the paper, details of the experimental design are crucial if statistics are used. In engineering this division is the heart of your paper, so the details are vital. Moreover, differences in approach limit the use of sentence patterns from previous writing.

> Writing details of the current research as you do the work is easier than trying to remember them all later.

For either format, because of the importance of details I suggest writing this division as you do your work. It is much easier to remember and write details as you do them than several months later. Rewriting is easier than recalling. Begin by writing more details than you think necessary. What seems obvious to you will not be clear to others who cannot see your lab setup.

One of the greatest struggles for students writing in IMRD format is deciding what to include in the Results, what to place in the Discussion, and how to connect the two. This struggle is reflected in research and textbooks on academic writing, which show substantial overlap in what they say should be included in each section.[2] Traditionally, data should be placed in the Results section, with comparisons and interpretations in the Discussion section. In principle this separates the evidence, considered to be correct, from the interpretation, which needs to be verified.[3] Nevertheless, since this places the evidence and reasoning in separate sections, neither can stand as a complete argument. Writing a combined Results and Discussion division and laying out the basic structure of the argument first will help (section 19.2).

10.3 Goal, audience, criteria

In any writing it is crucial to consider three things: your goal, the audience, and the criteria by which it will be judged. Although a conference paper, a journal article, and a thesis may have similar content, they are quite different in structure and detail because of those three factors, so you need to write with a purpose in mind. As the American novelist Carl Hiaasen said in an interview, "You can do the best research and be making the strongest intellectual argument, but if readers don't get past the third paragraph you've wasted your energy and valuable ink" (in Ferguson, 2002).

> Who will read your work, why, and most importantly, what criteria will they use to judge whether it is good?

Most things we do have both an immediate and long-term goal, or at least short- and long-term impact. Obviously, if you are a student your immediate purpose is to graduate, and certain writing tasks will be required for that, but what is your long-term goal? If you are looking for a job in academics, you may choose to write differently than if you want to go into industry, focusing more on theoretical aspects instead of practical. Writing not only showcases your research, but makes you visible, for better or worse.

Next, how large is your audience, how much do they know about your topic, and why do they care about it? The primary audience for a thesis or dissertation is your advisor and committee. They already know quite a bit about your research and you can guarantee they will read it carefully because that is their job. A conference paper or journal article may be addressed to a group of specialists or a broader audience, but you need to attract and keep their attention if you want your work to be read and cited. This will affect how much background you need to provide and how much detail you can or need to include.

Perhaps the most important consideration, often overlooked by first-time writers, is the criteria by which the work will be judged. A successful thesis must attain a certain standard set by your department. A conference paper must fit the theme of the conference before reviewers will consider the quality of the content. A research article is judged by many criteria including whether it fits the scope of the journal, the breadth and currency of the references, the novelty and importance of the work, the organization and quality of writing, and the clarity and necessity of graphics.[4]

Sometimes you have at least two separate audiences, reviewers and readers, with different criteria. If you do not convince the first audience that your work has merit, the second audience will never see it. The former will likely be quite aware of previous work in your specialty and only want to see how your work builds on it. On the other hand, some of the second audience will be graduate students with a less extensive background who need more details.

Whatever you write, do the best you can. Most work now is in electronic format, making it available to others far beyond your immediate audience, so it is important to make sure it represents you well.

Evaluate your understanding

(1) How can adding a one-sentence summary of an article to a document you are writing improve your memory of that article?
(2) Why is it important to write the details of your research as you do the work?
(3) Why should you consider your goals, audience(s), and the criteria by which your work will be judged?

Homework

(1) What do you plan to write first?

Hint: See the following chapters for various types of writing. Read whichever is appropriate to your current writing, then skip to Chapter 18 to begin writing. Come back to the others as needed.

(2) What is the goal of your writing, both short and long term?
(3) Who is the audience for your writing? How much do they know about your topic? Why are they or why should they be interested in it?
(4) By what criteria will it be judged?

Hint: Some of this information will be available from the instructions to authors (journal articles), call for papers (conference papers), department regulations (theses), or professors (undergraduate writing). Some you may need to obtain by talking to others who have successfully completed the task you are undertaking. Keep these questions in mind as you write, and more importantly as you rewrite, deciding what to cut or add.

Teaching tip: Many of the ideas presented here will be developed further in subsequent chapters. You may find it useful to discuss how these principles relate to the situation and type of writing students are doing in your class.

Notes

1 Hyland has written extensively on stance, e.g., Hyland (1999), Hyland and Guinda (2012).
2 Examples will be given in Rau (In preparation).
3 The distinction between evidence and inference is not always clear. See, for example, the discussion of indirect evidence in Rau (2012: 60, footnote 7).
4 See Chapter 27 of this text and Cargill and O'Connor (2013: 95).

References

CARGILL, M. & O'CONNOR, P. 2013. *Writing scientific research articles: Strategy and steps*, New York, John Wiley & Sons.
FERGUSON, E. 2002. He's all the rage. *The Guardian*, 17 March.
HYLAND, K. 1999. Disciplinary discourses: Writer stance in research articles. *In*: CANDLIN, C. & HYLAND, K. (eds.) *Writing: Texts, processes and practice*, Harlow, Longman.
HYLAND, K. & GUINDA, C. S. 2012. *Stance and voice in written academic genres*, New York, Springer.
RAU, G. 2012. *Mapping the origins debate: Six models of the beginning of everything*, Downers Grove, IL, InterVarsity Press.
RAU, G. In preparation. IPTC: A prototypical format for engineering research articles.

Chapter 11

Undergraduate writing

11.1 So near, and yet so far

When climbing a mountain you may rejoice as you surmount a peak, before looking up to see the real summit looming above you in the distance. Similarly, having mastered undergraduate writing you think you have arrived, only to find that there are many higher peaks beyond. Nevertheless, if you have strengthened your skill in writing arguments, you will be prepared for the path ahead.

We will begin our look at various genres with three types of documents common in undergraduate courses. Naming them is a challenge, as the terms used differ by country, institution, department, and course, but we will call them lab reports, essays, and senior projects. For each we will consider the goals, audience, and criteria, and how those affect the structure of the document.

On the surface each genre may appear to closely resemble research articles, but at a deeper level differences appear. Lab reports often have the same section titles but lack many components found in research articles. Essays are usually secondary research, more similar in structure to survey or review articles. Experience writing either of these may contribute to difficulties in later writing if you are not aware of the differences. Senior projects, particularly those involving primary research, are longer term, offer more autonomy, and thus may be like graduate theses.

> Undergraduate writing lacks many components of research articles, but practice making and supporting claims will prepare you for future writing.

Undergraduate writing tasks allow you to build confidence as a researcher and a writer. Attention to argument structure, making and supporting claims, will help you in any future writing. Senior projects may be written up for submission to a conference or journal, and beneficial for graduate school or job applications.

11.2 Lab reports

Lab reports are perhaps the most common undergraduate writing task for science and engineering students. Classroom laboratory experiences are usually demonstrations of students' ability to follow procedures, use apparatus, and obtain, present, and interpret results, but not part of an ongoing program of authentic research. Although instructors guide students to use section titles reflecting research articles in the field, the similarity is only skin deep, because *the implicit claim is that you have mastered certain methods or principles, not that you are contributing to knowledge.*

First we need to distinguish three types of laboratory experience. Many traditional laboratory exercises follow a "cookbook" approach, where the exact procedure is specified. Increasingly, science education is encouraging use of an inquiry model, where students must formulate a question and find a way to answer it. Engineering or technical courses may use a design challenge format (Rau & Antink-Meyer, 2020).

In the cookbook approach students must follow the procedure and can predict the expected results. Sometimes this leads to falsification of data (*I* wouldn't have any personal experience with *that*!), but that is not the issue here. What is important for our purposes is that this is very different from actual research. There is no reason to include Component 1 ("This is important because the instructor said to do it.") or Component 2 ("This has already been done thousands of times with the same result, but here we go again."). Usually the goal is stated ("We will verify the natural law that has been verified with far greater precision for centuries."). There is no need to explain the framework or the details of the method, as they are already listed in the lab manual, although usually some summary is required. So what is left? The data and explanation. While it is useful to produce and explain tables and graphs, that is by no means the same as writing a research article. If you are asked to write up a cookbook lab, you already know how to get exemplars (from former students who took the course), and your report will be very similar to theirs (hopefully not *too* similar), so there is no need to talk further about this.[1]

Inquiry and design challenges are more like actual research, but with differences that can be attributed to the audience, goal, and criteria. *The audience for all student work is the professor, who also sets the goal and criteria.* The professor and former students are the only possible sources for exemplars for that specific class. Since each professor values different things, you will have to look closely at the stated criteria (or infer the unstated from the way they teach and grade) to succeed.

> **Lab reports often lack Components 1, 2, and 4, modify several others, and have little or no support from citations of past research.**

Nonetheless, a few general comments can be made about what components should be included.[2] Some sort of introduction is necessary to establish the topic of the report. This may include importance but frequently only provides background information. Citations may be required for inquiry labs. The goal should also be stated in the introduction, but it is the goal of the lab, not a research goal. The procedures used are stated next, but materials supplied by the instructor may not be. Statistical procedures, if used, should be reported. The data are shown, often as tables or graphs, with interpretation, including comparison with predictions. A separate conclusion is common in engineering lab reports, but not science, as in the respective research articles, but is a summary of the lab rather than a claim of contribution.

The most noticeable absences are the importance, need, and framework, along with citations to support them. Lack of experience with this may make it difficult to appreciate the importance of connection with past research and write about it when doing actual research.

11.3 Essays

The second main genre of undergraduate writing is the essay, also called a report or research paper. It reports previous work rather than original research. Thus the format is closer to a survey or review article, or secondary research in social science, than to primary research articles in science or engineering. Essays seem to be more common in science than engineering.

As with lab reports, *the professor is the audience, and the assignment should state the goal and criteria.* Usually the *goal is to attain a deeper understanding of one topic related to the class.* Exemplars are again specific to each class, although the structure of review articles may provide helpful guidelines (section 13.4).

> Essays often lack Components 2, 5, and 6, and modify several others.

The introduction of an essay will focus only on the importance of the topic, as there is no reason to prove the need for your paper. A statement of purpose or research questions is required. The methods you followed to collect your information are not needed and since no new testing was performed, there are no testing methods. Instead, various frameworks for approaching the topic may be discussed in detail. Specific results from the literature may be included to illustrate or focus attention on key points. By far the most important portion is the discussion, where, depending on the assignment, you may be asked to summarize, compare, evaluate, or integrate, so what needs to be included will vary accordingly. Generally, references and citations are required.

Since the requirements for an essay are quite varied and dependent on the individual professor and goals of the course, it is hard to generalize about what to include. Nevertheless, argument structure is crucial. What claims are you making? Your first claim is probably the importance of the topic. Next you might claim there are three main approaches. For each approach you might need to document, with examples, their strengths and weaknesses. Finally, you claim one approach seems to offer the most potential. *The claims will differ depending on your topic,* so think about the argument structure and support before you begin writing (Chapter 19). With secondary research, there are many options for sequencing, especially in the discussion (section 9.5). As you rewrite, consider alternative orders that would strengthen the argument or make it clearer for the reader.

Essays differ from review articles and research writing in one other key aspect. In an essay you are providing a summary of a topic, but need not include all related work. A sampling or selection of some of the most important work is usually sufficient. However, to prove that there is a need for your research, you must show you are familiar with all the work in the field, including the most recent, to show there is still a gap. All graduate writing and review articles require a comprehensive literature review, so this is an important difference to keep in mind in later writing.

11.4 Senior project

This genre is very departmentally dependent. In some departments all seniors majoring in a department write a final paper as part of a senior seminar. In others it is an option for honors. Thus the document may be called a senior capstone, senior thesis, honors project, and undoubtedly many other names.

> Senior projects requiring original research are like graduate theses; others are like essays.

Senior seminars may not require original research, in which case the paper will follow the essay genre (section 11.3). Honors projects in science and engineering usually require original research, so the writeup is more like a graduate thesis (section 12.6), but with lower expectations for length, number of references, and depth of research.

If you are writing a senior project, the first step is to get both the requirements (criteria) and exemplars from your department. Sometimes the project will be read only by one professor, sometimes by several. It may or may not require an oral presentation. For a seminar you may have complete freedom of choice in essay topic. Original research often must be related to the work of an advisor, because of restrictions of budget and material.

11.5 Summary

Table 11.1 summarizes which components tend be found in each type of undergraduate writing, and differences from research articles. One of the main differences is that since the goal in lab reports and essays is not original research, it is not necessary to show the need for the work or the connection with previous research. Furthermore, the methods and testing methods are less important than the results and discussion. Senior projects that

Table 11.1 Components and support in undergraduate writing

Component/support	*Lab reports*	*Essays*	*Senior project*
1 Importance	No	Yes	Yes
2 Need	No	No	Yes
3 Research goal	Goal, not research	Goal, not research	Yes
4 Framework	No	Yes	Yes
5 Research details	Summary	No	Yes
6 Testing methods	Yes	No	Yes
7 Data patterns	Yes	Examples	Yes
8 Comparisons	Yes	Yes	Yes
9 Interpretations	Yes	Yes	Yes
10 Conclusion	Summary	Summary	Yes
I Past research	Little or none	Yes	Yes
II Present research	Less detail	No	Yes
III Logical	Usually deductive	Usually inductive	Yes
IV Evaluative	No	Yes	Yes
V Organization	No	Yes	Yes

involve original research will contain most if not all of the components found in research articles. Remember that general summaries may not apply to your situation, since these genres are often assigned and graded by a single professor.

Evaluate your understanding

(1) How is each genre of undergraduate writing like and unlike research articles?

Homework

(1) Obtain exemplars and criteria for whatever you plan to write.
(2) Before you start to write, decide what claims and components you will need to include and how you will support the claims.

Hint: Plan and write following the steps in Part 4, beginning in Chapter 18.

Teaching tip: Help students obtain suitable exemplars for whatever you want them to write and compare them with the structure of research articles.

Teaching tip: This chapter may be useful for graduate students as a way of contrasting their current writing with formats they have written previously.

Notes

1 Obviously, an author's evaluation does not need to be stated explicitly to be clear.
2 See Parkinson (2017).

References

PARKINSON, J. 2017. The student laboratory report genre: A genre analysis. *English for Specific Purposes*, 45, 1–13.

RAU, G. & ANTINK-MEYER, A. 2020. Distinguishing between science, engineering and technology. *In*: MCCOMAS, W. & ORAMOUS, J. (eds.) *Nature of science in science instruction: Rationales and strategies*, New York, Springer.

Chapter 12

Graduate writing

12.1 Apprenticeship

Graduate work is an academic apprenticeship. Working with your advisor and older graduate students, you slowly learn to do research. Reading other research, you gradually realize what characterizes both good research and good writing, which do not always go together.

As the American author William Faulkner said,

> Read, read, read. Read everything—trash, classics, good and bad, and see how they do it. Just like a carpenter who works as an apprentice and studies the master. Read! You'll absorb it. Then write. If it's good, you'll find out. If it's not, throw it out of the window.
>
> (in Meriwether & Millgate, 1968: 55)

As you begin your own writing you will find that some of it is trash. Throw it out the window (being careful not to hit anyone passing by) and start over. Like a carpenter, start with straight cuts and common nails, doing the rough framing that is hidden once a wall is completed. Gradually move to bevel cuts and finishing nails on the surface that shows. Eventually you may be ready to do dovetails without nails, the neat and smooth work of a cabinetmaker. But you cannot get there without practice.

Graduate writing includes most of the components, but to differing degrees.

As a graduate student you may continue to write lab reports and essays for classes, described in the previous chapter. You will also have to write at least three types of documents related to your research: research articles, conference papers, and your thesis or dissertation. Engineers may write for industrial journals. Your department may require a research proposal. All include the same components we have studied, but with different emphasis.

12.2 Research articles

Research articles are the main publication for which you get recognition in academia. Since we have already studied their structure, this section will briefly mention a few things to keep in mind as you write.

Always consider goal, audience, and criteria (section 10.3). Each journal has a different audience and therefore expectations. A selective, indexed journal and a more specialized one have different criteria, and you may have a different goal in submitting to each (Chapter 27). Some journals require payment for publication. These often accept almost every paper submitted, with minimal review, and therefore get a reputation for publishing low-quality work. Steer clear of such journals.

> Publishing the same information in two places is considered plagiarism, regardless of the language of publication.

You cannot submit the same paper or publish the same information in two places. How much difference is enough? Some authors boost their number of publications by making only one minor modification to each publication. This may pad your resume but will be annoying to other researchers. For example, as a graduate student I found two papers, published by the same author, the same year, in two different journals. One concluded that a method of genetic analysis was useful, the other that it was not. Careful comparison showed that although the wording was very different, both were based on data from the same plants, the same years, but examining different traits. Neither article mentioned the other, so future researchers (like me) were left searching through reams of information to determine why the conclusions were so different. Needless to say, I do not recommend that technique.

For those working in English as a lingua franca, it is also important to emphasize that publishing a translation of the same paper is considered plagiarism, since the content remains the same. A translation must be designated as such, with permission from the original publisher, who holds the copyright. Usually it is better to do further research and publish a subsequent paper in the other language that cites the original.

12.3 Research proposals

When required, a research proposal is the first document you will write related to your graduate research. A formal research proposal may not be required for a Master's degree, as the topic is often assigned by the advisor, a small part of some larger project. It is a common but not universal requirement for a doctoral degree program. Usually the proposal will be read by several faculty members. It often requires an oral presentation and defense of the proposal, at which the committee will ask questions for clarification or make suggestions for changes. *The implicit claim you need to support is that you are prepared to do the research.*

The purpose of the proposal is to ensure that you are familiar with other research in the field, have considered what topics are important but need further work, are aware of various methods used to answer such questions or design a solution, and have the ability to complete the research. In some fields it is like an extended Introduction, but in others requires additional components not found in research articles. As with any writing, the first step is to determine the requirements of your department and, if possible, obtain exemplars of previous proposals to determine unstated expectations such as length and degree of detail.

Before writing a research proposal you will need to have a general topic and read extensively, probably at least 50 and perhaps over 100 articles for doctoral work. Not all the articles you read will turn out to be directly related to your final project, so not all may be listed in your final proposal, but from the beginning it will be worthwhile to use a reference manager to keep track of everything (section 25.3). As you read, you will find out what are currently considered important questions or problems in your field and what has already been done. Your proposal must build on what has already been done, adding to it in a unique and substantial way. You may need to do a pilot study to demonstrate the potential of your method.

Research proposals frequently include Components 1, 2, 3, 4, 5, and 6, and may require estimates of cost and time.

Departmental policies will guide your overall organization. If none is specified in the regulations, you may be able to discern an expected structure from exemplars. The section names and order may differ, but the following topics are commonly found in research proposals:

- Objective (research goal);
- Introduction (importance, need, previous work);
- Framework (alternative approaches, justification);
- Procedures (methods or anticipated design, testing methods);
- Resources needed (materials, cost);
- Timeframe.

Unlike a research article, the research objective may be stated at the beginning to make the proposal easier to evaluate. If the goal is not clear, too close to current work, or unattainable, there is no need to read further. Thus it is important to discuss this with your advisor before writing.

As in a research article, the Introduction will often include both the importance of the topic and the need for further work, each supported by citation of previous work in the field. The further the proposal is from current work in the field, the more justification will be required. This is frequently separated from a longer description of previous research, again for ease of evaluation.

The third part may have many names, including literature review, background, related work, or framework. It includes extensive evaluation of previous research or approaches, their strengths and limitations. It frequently includes a justification for following or building on one framework.

This is followed by the anticipated methods (science) or design (engineering), and testing methods. Although these may change during the research, a full description is required, including complete experimental design if statistics are used. If the method is not feasible or the testing will not produce meaningful results, it is better to find that out before beginning. The anticipated results or predictions usually are required, along with results of any pilot studies or prototypes.

In addition to these topics, which mirror the first six components, many proposals also require an estimate of the resources needed, their cost, and a timeline. Although these are likely to change, it helps ensure that the project can be completed given the time and resources available, or perhaps point to the need to write a grant to cover the cost.

A research proposal could be anywhere from 10 to 50 pages, so again it is important to determine the expectations of your department. Most of what you write will be included, after many additions, changes, and revisions, in your thesis or dissertation and other publications.

12.4 Conference papers

Often the first document you will write for an audience outside your department is a conference paper. On the surface conference papers appear identical in structure to research articles in the same field, but there are differences in goal, audience, and criteria. They have a narrower scope and are less complete and less carefully reviewed, and thus have a lower impact than journal articles.

Conference papers give you visibility in the field. Some conferences are highly selective; others not at all.

The goal of writing a conference paper is to attend the conference. Attending allows you to meet other researchers in the field, gain recognition for yourself and your work, and hear about the latest developments in the field. Usually to get funding to attend you must present either a paper (section 16.4) or poster (section 16.6). Presenting is a good opportunity to practice your speaking skills before your thesis defense or job interview.

There are *several audiences for your conference paper.* At a minimum the paper will be seen by reviewers evaluating its suitability for the conference. If accepted, it will be available to participants in the conference. Some conferences have graduate student paper competitions, a third audience. Particularly in engineering, conference papers are frequently published as Proceedings, making them available to a far broader audience.

When submitting a proposal to a conference, consider what information will be available to each audience and when. You want people to attend your presentation, not just read your paper. Will they have the full paper or only the abstract available before the conference? Your abstract should attract attention to both.

Criteria for acceptance of conference papers vary. Some conferences are very selective, and even experienced researchers can have papers rejected. Others accept almost every submission, leading to thousands of participants and multiple parallel sessions. Conferences often ask that submissions be related to a certain theme or track. This facilitates screening proposals, as a different group of reviewers is assigned to each. If you submit to the wrong track your work is less likely to be accepted. In a non-selective conference, a paper submitted to the wrong track may be accepted, but not seen by interested participants.

Conference papers normally have a narrow focus, present only preliminary results, and are not rewritten based on reviewer comments.

The scope of a conference paper is often narrower than a research article, being a written record of what you will present orally in less than 20 minutes. Instead of the whole design, you might focus on one element, or only one of several traits you are testing. This also allows you to present the same project at different conferences, since it is considered improper to repeat the same information in two venues.

Conference papers often must be submitted six months or more before the conference, and therefore report only preliminary results, with the expectation that further results will be presented verbally at the conference. Thus the components of the Results and Discussion are often missing or very short relative to research articles in the same field. Nevertheless, the *implicit claim is that you will have something worth listening to*.

Because the same term is used, novice writers may not recognize the difference between review of conference papers and journal articles. Review for a conference is limited to checking the suitability for the conference and the quality of the work, but rarely are reviewer comments given to the author, with the chance to revise and resubmit. Thus conference papers frequently have less complete citations of past research and less detailed descriptions of the current work than research articles.

For all these reasons, conference papers are not considered full publication, and have less impact than research articles. Nevertheless, conference papers are important in engineering. Since they report the latest developments in a field, conference papers are frequently cited in engineering research articles. In science, with its emphasis on data and analysis, citation of conference papers is extremely rare.

> **Conference papers are cited frequently in engineering, but rarely in science.**

12.5 Industrial journal articles

In some fields of engineering, graduate students may write articles in journals with an industrial audience.[1] Whether you choose to write for these journals or others depends in part on whether your long-term goal is work in academia or industry. The underlying *claim is that the technique presented has immediate practical utility in industry*, so although the overall structure is similar to IPTC, some components are minimized.

The Introduction usually begins with the specific importance to industry, rather than society at large or researchers in the field. Need is similarly limited to problems with the current best solution, with less discussion of historical or alternative methods and therefore fewer citations. The research goal is often in a separate section or subsection.

Framework is introduced only as it relates to the current solution, which is presented at length. Testing emphasizes the advantage of the present method, often with less attention to comparison with other methods and more details on specific numerical test results, so readers can compare it with their own current method or industry standards.

Overall, the articles may be shorter than other research articles, but have as many or more graphics. They frequently lack organizational or summary paragraphs, although topic sentences are prominent. These generalities would of course need to be confirmed for the journals in your field.

12.6 Thesis or dissertation

First it is necessary to clarify terminology. In American usage, a Master's student writes a thesis, while a doctoral student writes a dissertation. The opposite is true in British English. For ease, in this section I will use the term thesis to refer to either.

The structure of a thesis is usually similar to a research article, but longer. What would be sections in an article become chapters in the thesis. The thesis may include material that will eventually be published in several articles, as well as some that will never get published, including extensive appendices, data tables, or approaches tried that did not work or produced inconclusive results.

Like a research article, a thesis begins with an Introduction. In some fields this will follow the same structure as a research article, while in others it is a synopsis of the thesis, beginning with the research goal, followed by a summary of the methods, results, discussion, and conclusions.

Frequently the second chapter is a literature review, far more extensive than will be published in any of the related articles. The purpose of this is to demonstrate to the committee that the student has a clear understanding of various approaches to the topic, not only the one used in the thesis.

> The thesis may be organized like an extended article or a compilation of several articles, preceded by a general Introduction and Literature Review.

Beyond this, the thesis will follow one of two nested designs. The first is more like an extended research article, with chapters designated Methods, Results, Discussion, and Conclusion. In this case within each of the three main chapters there will be sections for different experiments, discussed in the same order in each chapter. This works well for closely related experiments with similar methodology.

In other cases, the chapters will be organized more like a series of articles, each with Methods, Results, and Discussion (or appropriate sections in IPTC format). This works better when more diverse methods or designs have been used. It may also be chosen if some of the material has been or is in the process of being written as articles, to minimize the need for rewriting and make it easier for readers to relate the two.

In either case, the Conclusion is likely to be longer than the one in a research article. Frequently it includes a summary of each chapter before drawing overall conclusions. The abstract is also more likely to be a summary of the entire thesis rather than an advertisement highlighting the main contribution only.

Criteria for the contents and length of a thesis are set by each department. The format may be set by the university. If the thesis is to be made available online, as most are currently, there may be additional requirements for that. All the necessary information should be available from the department secretary or a thesis secretary in the graduate school office. Often there are several stages of approval, even after the content has been approved by your advisor and committee, so make sure you check the requirements and allow sufficient time to meet them before the deadline.

12.7 Summary

The main genres associated with graduate research are all closely related to research articles but will be judged by different criteria. As shown in Table 12.1, the research proposal is preliminary, presenting the planned methods and expected results, both of which will undoubtedly be modified in the course of the research. A conference paper usually contains at least preliminary results and may be cited in engineering before a related article is published. Industrial journal articles are more practical, lacking or minimizing some of the components. A thesis (dissertation) is longer than several articles and summarizes all the work done, whether publishable in another format or not. It frequently includes more detail on both the method and results than any of the other genres, as well as a far more extensive literature review.

Evaluate your understanding

(1) How does each of the other genres of graduate writing differ from research articles?

Homework

(1) Obtain exemplars and criteria for whatever you plan to write.
(2) Write as you do the work, then revise, following the suggestions for research articles.

Teaching tip: Help students find appropriate exemplars for each genre.

Table 12.1 Components, support, and additional items in graduate writing

Component/Support	*Research proposal*	*Conference papers*	*Industrial journals*	*Thesis or dissertation*
1 Importance	Yes	Yes	Practical	Yes
2 Need	Yes	Yes	Yes	Yes
3 Research goal	Yes	Yes	Yes	Yes
4 Framework	Yes	Short	Maybe	Yes
5 Research details	Planned	Yes	Yes	Yes
6 Testing methods	Planned	Yes	Yes	Yes
7 Data patterns	Expected	Preliminary	Yes	Yes
8 Comparisons	Expected	Preliminary	Yes	Yes
9 Interpretations	No	Preliminary	Maybe	Yes
10 Conclusion	No	Preliminary	Yes	Yes
I Past research	Yes	Some	Some	Yes
II Present research	No	Preliminary	Yes	Yes
III Logical	Yes	Yes	Yes	Yes
IV Evaluative	Yes	Maybe	Yes	Yes
V Organization	No	No	No	Yes
A Resources	Yes	-	-	-
B Timeframe	Yes	-	-	-
C Unpublished data	-	-	-	Yes

Research articles (not shown) contain Components 1–10 and support I–V.

Note

1 At our school this is common in mechanical engineering, but not electrical, chemical, or computer science.

Reference

MERIWETHER, J. B. & MILLGATE, M. 1968. *Lion in the garden: Interviews with William Faulkner, 1926–1962*, New York, Random House.

Chapter 13

Academic writing

13.1 Never stop learning

If you remain in academia you never stop learning, whether about your field or about different genres of writing. You will continue to write conference papers and research articles but will also branch out into other genres. At a minimum you will have to write grant proposals, and probably other types of articles, book chapters, or even books. Unlike the genres presented in the previous chapter, these are all quite different from primary research articles. Nonetheless, they contain many of the same components and make the same basic claim that you are contributing to knowledge.

More time will be spent discussing grant proposals, since they are the most distinct. You will probably discover how the others differ from research articles in the course of your own reading.

13.2 Grant proposals

Grant applications are difficult because they require skills in management and finance, rarely taught in science and engineering programs. Grants involve not only the initial proposal, but also progress reports on the funded research (section 14.2). Doctoral students frequently have a hand in writing a grant proposal to fund their research or the ongoing research program of a lab. Although similar to research proposals, the stakes are much higher. A failed grant proposal may delay your work.

> For grant proposals it is essential to follow the guidelines, and helpful to examine successful past proposals or apply with an experienced researcher.

Every granting agency has different formats, requirements, and criteria for grant proposals, so it is important to both check all relevant online documents and examine examples of successful proposals. Writing your first grant as a co-PI (principal investigator) with an experienced researcher can help you learn the format and increase your chances of success. In general, *the main claim is that the proposed research is important to the grantor and that your lab has the ability to complete it successfully*. Grant proposals include at least:

- Applicant and institution contact information;
- Justification (pursuant to grant goals);
- Framework and related work;
- Research goals;
- Experimental design or research plan;
- Expected findings;
- Participant expertise and institutional support;
- Timeline;
- Budget.

Grant proposals require most of the same components found in research articles, although stated as plans or expectations rather than accomplished fact and not always in the same order. Additionally, estimates of budget, timeframe, and an evaluation of the resources available to complete the work are required. Frequently the granting agency will provide an electronic form to fill in, often with a page or word limit for each category. Failure to comply with those guidelines will lead to swift and certain rejection.

Grant proposals begin with several pages of forms, providing contact information of the researchers and institution. If your research involves animal or human subjects, additional questions will ensure compliance with guidelines for ethical research and approval from an Institutional Review Board (IRB).

In the written portion, every grant application must show the importance of the research, particularly how it meets the goals of the granting agency. Contribution to society or future applications are secondary, unless these are stated goals of the agency.

A review and evaluation of related work is also required, frequently in a separate section. This need not be comprehensive but must include the most recent related work, particularly ongoing work in your lab. Like any other research, the application must build on an existing model or framework, except grants for creative solutions, which require more documentation and reasoning to show their feasibility. Being able to explain the framework clearly and concisely shows your familiarity with the concepts, thus may be more prominent than in research articles.

Instead of a single research goal, grant applications typically state three goals: easy, achievable, and challenging (Poe et al., 2010: 88). The first should be something that will be completed by the end of the first reporting period, often within the first six months or year of the grant. This allows you to point to progress in your first report. The second is something you are reasonably certain you will be able to achieve by the end of the grant. This allows you to claim success in your final report. The third is something that you expect to work on but not necessarily complete during the term of the grant. This lays the basis for extensions or future grants.

The experimental design or research plan is also crucial, as it allows the grant committee to evaluate the potential for success of the work. This should be as detailed as possible within the length guidelines of the application.

Later components of a research article are included as plans or expectations. Expected results, discussion, and conclusion should be stated confidently, but with appropriate qualification and hedges (Chapter 22). The goal must be measurable and the method for evaluating success clear.

Beyond the components of a research article, you will need to demonstrate competence and ability to complete the research, both with personnel and facilities. Core continuing personnel are listed not by name but by experience, including publications, awards, and familiarity with specific protocols. Personnel to be brought on later, including graduate students whose research will be funded, do not have to be listed, but the criteria of selection may be. You must show that the lab, department, or school has enough basic resources to manage and complete the research, or that you will partner with other organizations with the necessary expertise or facilities.

Budget and timeline must also be included. Underestimating or overestimating either can lead to failure. Of course, both must be within the guidelines set for the grant. This is one of the areas where it is most useful to look at previous grants, to ensure that you have not forgotten any categories of expenses or severely misjudged cost or time.

13.3 Short papers

As mentioned in Chapter 2, journals may contain genres other than primary research articles. One of the more common is short papers, brief reports, letters, or correspondence. These are common in rapidly changing areas of technology, providing an opportunity for researchers to publish quickly, with minimal editorial review, largely in order to claim primacy for a discovery or new design. The practice of noting the dates a document was received by the editor, accepted for publication, or published online prior to print publication serves the same purpose.

> **Short papers may have simple organization and minimize many components.**

Generalization about the format of short papers is difficult, as there are many variations. Some are several pages long, with a structure nearly identical to journal articles, including four sections representing IMRaDC and most if not all components, but fewer details and references. Others may be only one or two pages with no sections and lack several components. Minimal review means the structure is determined by what the author finds essential to the *implicit claim that this is worth publishing quickly.*

The Introduction is usually shorter than a research article. There may be only a short statement of importance or none. Discussion of previous research may be limited, or many papers cited but with minimal evaluation. The research goal may be stated at length, comprising half of the Introduction, or in a single sentence. Framework and details of the current research are often minimized, mentioning only novel features, but may be very important in engineering.

The focus of the article is often on the Results and Discussion (science) or Testing (engineering), with sufficient background to understand the cause or implication of those results. Graphics are often present, but with minimal commentary.

Some journals have a relatively fixed style for short papers, while others allow far more flexibility. Doing component analysis of a few articles from your target journal will allow you to figure out the expectations and write accordingly.

13.4 Review or survey articles

The word "review" can be used as a verb (Chapter 18, note 2), a noun (blind review, Chapter 27; book or product reviews), or an adjective (review article). In this section we are talking about the latter: articles written to review or survey the status of a field.

Review articles may be written by an experienced researcher as a way of setting direction for the future or assigned to a doctoral candidate to help them map the territory and find a research topic. Since they represent the opinion of one researcher, they are sometimes called a Perspective.

> Review articles do not report a new design or results, and therefore lack many components.

Since review articles do not report new research, they lack many components, including the research goal (although the goal of the review will be stated), current research details, and testing methods. While data from other research may be reported and compared, no new data are reported, so there is no Results or Testing division. On the other hand, there is extensive evaluation of alternative frameworks and interpretation of various results.

While the section structure differs, the order of review articles is relatively constant. They typically begin with a more extensive discussion of Importance than research articles, in some cases comprising a whole section. Then a summary or grouping of various approaches precedes a detailed evaluation of each. Finally, the various approaches are compared.

Some reviews are comprehensive, mentioning every paper published on a topic within a specified time period, while others are broader, highlighting major papers and shifts in a field. They may be presented chronologically or follow some other organization scheme (section 9.5). Results obtained by different methods may be compared graphically or only in text, but interpretation is expected.

Before writing a review article, *make sure you are clear about its purpose.* If the main purpose is to help you find a researchable topic, it will help others do the same. This requires extensive evaluation of the advantages and drawbacks of each approach. Whether those can be identified from the methods themselves or the results obtained, from rational evaluation or empirical comparison, depends on the topic. If the purpose is to point to new directions for future research, you may wish to emphasize the diminishing returns of current methods and the foundations and potential of new approaches.

13.5 Book chapters

Let me begin this section with a word of warning for those considering publishing a book chapter. Maybe two words of warning. First, it is less likely to be cited than an article (unless it becomes a hugely popular book). Journals are more likely to be available through your university library than a book, and no one wants to track a book down through interlibrary loan or pay the price of downloading a chapter. Second, don't count on a book chapter for your resume, let alone to report the latest advance, as it may take several years to be published. On the other hand, book chapters can be useful for introducing yourself and your research to a different group of specialists, particularly in crossover areas.

Book chapters seem to be less common in science and engineering than in social sciences and humanities, probably largely because of the importance of speed of publication. They may be important in areas where there is an overlap between technology and society, science and education, or newly emerging fields on the borders of previous disciplines.

Book chapters are usually solicited by an editor and must fit the focus and perspective of the book. How much control the author has over the contents depends on the editor. Some books are collections of research on one topic, so each article is very similar to a research article. Others are intended as primers, with each article more like a review article on one topic. *Make sure you understand the purpose of the book as a whole and how your chapter contributes to it.*

> **Book chapters in an edited volume may be like either research articles or review articles.**

For book chapters, contributors are usually chosen by the editor, and review is often done internally, with each author commenting on two other chapters, with additional comments from the editor. Since chapters are more likely to be undertaken by experienced authors, I will not say more on these.

13.6 Popular writing

While many scientists and engineers are content to communicate only with other specialists, others want to get their message out to a broader audience. Different genres of popular writing include science magazines, science teacher magazines, newspaper columns, online news services, blogs, social media, and many others. *Each genre has a different audience, expectations, and format.*

Some science and engineering organizations have popular journals as well as research and industry publications. Such journals can be distinguished by prominent use of colorful eye-catching visuals—the work of graphic designers rather than the author. Articles may be written by science writers, rather than research scientists, although prominent scientists may also be asked to contribute articles.

> **The purpose of popular writing is to attract the attention of non-specialists, employing graphic design and one main point, with limited details or qualifiers.**

Although the format of each varies, *the implicit claim is that this is something interesting and important to non-scientists or scientists from other fields.* Instead of a general statement of importance, the article will begin with an attention-catching example, illustration, metaphor, or statistic. Each article normally focuses on one small topic, sometimes not even the main point of the original research article it is drawn from, with minimal mention of other research and few if any citations. The interpretation of the results is often presented as truth, without either the data or qualifiers common in research journals.

Since popular writing encompasses many different genres, it will be necessary to do an analysis of several exemplars to determine the format and level of the audience. In many cases the publisher exerts far more control over the product than in a research article, even rewriting the title to make it more interesting.

Table 13.1 Components, support, and additional items in other academic writing

Component/support	*Grant proposal*	*Short paper*	*Review/ survey*	*Popular writing*
1 Importance	Yes	Maybe	Yes	Interest
2 Need	Yes	Maybe	Yes	No
3 Research goal	Yes	Yes	Goal	No
4 Framework	Yes	Short	Multiple	No
5 Research details	Planned	Short	No	No
6 Testing methods	Planned	Short	No	No
7 Data patterns	Expected	Yes	Previous	Summary
8 Comparisons	Expected	Yes	Yes	No
9 Interpretations	Expected	Yes	Yes	Yes
10 Conclusion	Expected	Yes	Yes	No
I Past research	Yes	Limited	Yes	No
II Present research	Planned	Short	No	No
III Logical	Yes	Yes	Yes	No
IV Evaluative	Yes	Maybe	Yes	No
V Organization	Forms	No	Yes	No
A Resources	Yes	-	-	-
B Timeframe	Yes	-	-	-
C Budget	Yes	-	-	-

Book chapters (not shown) may be like research or review articles.

13.7 Summary

All the genres in this chapter have a different structure than research articles, although some components overlap. To learn how to write each, obtain exemplars and do your own component analysis. Use Table 13.1 as a guide for the types of components common in each, but realize there is extensive variation. Book chapters, not shown in the table, may be similar either to research articles (with all the components and support) or review articles.

Evaluate your understanding

(1) How are these genres of academic writing like and unlike research articles?

Homework

(1) Obtain exemplars and criteria for whatever you plan to write.
(2) Talk with an experienced author of the genre you are writing, to get tips specific to your field and audience.

Reference

POE, M., LERNER, N. & CRAIG, J. 2010. *Learning to communicate in science and engineering: Case studies from MIT*, Cambridge, MA, MIT Press.

Chapter 14

Technical writing

14.1 Intertextuality

All academic and technical writing is part of an ongoing process. Your work is connected to other work that has been done before and currently by other workers. Thus no science or technical writing stands on its own, but always in reference to other writing. This connection between writing is called intertextuality (the relation between texts).

> Intertextuality makes it difficult to understand any text without the context.

While it is important in academia to show how your work connects with previous work, by citing past research, *in business and industry it is often assumed that those reading a document know the connection with what has gone before.* This makes it hard for a newcomer to join the conversation until they know the background.

This is especially true for the reports discussed in this chapter and email (Chapter 17), all of which build on previous reports, meetings, or correspondence, usually without specifically referring to it. Thus they would make no sense to outsiders, but they are not intended to. Imagine walking into the middle of a conversation between two friends about a recent sporting event. If you saw the event, you could join in. If you follow the sport, you could ask relevant questions. Otherwise it would mean nothing to you.

Scientists and engineers need to write various types of reports, the names of which differ by location, field, and industry. In the following sections I will discuss progress reports about an ongoing project, technical reports and management reports written for technical and non-technical audiences, respectively, and SWOT (Strengths, Weaknesses, Opportunities, Threats) analyses, used for a variety of purposes.

These types of reports are not public, making it harder to find exemplars. Each organization has a different preferred format. *Furthermore, because of intertextuality, authentic practice is not possible in a class situation.* Thus much of your learning about these genres must come on the job. Nevertheless, understanding the goal of each report type will help.

14.2 Progress reports

Progress reports share a *common central claim: that you have made progress on a project.* That project may be your graduate research, with an informal weekly report to your advisor or lab group. It could be a monthly project report in industry. It might be an annual progress report on a grant. In almost all cases, the *main audience is someone in a supervisory role*, with your colleagues as the secondary audience. In all cases, those involved are familiar with the work, so *the report only needs to highlight changes since the last report*, rather than the entire project.

Several factors influence the contents of a progress report, including the frequency, the stage of the project, and the relationship between the author and the recipient of the report. The frequency of reports may vary during a project, perhaps being more frequent during the early planning stage and as it nears completion but less frequent in the middle. Less frequent reports or relationships that are more distant require greater detail and formality. The length and format depend on the formality of the report. A weekly report to your advisor may be a list and some graphics on your computer. An annual report to a granting agency will be many pages long and include changes to personnel, process, or timeframe, and an accounting of funds.

Since the main claim is that you have made progress, the main support will come from your present work. Viewed in terms of components, that means the components represented are likely to be research details (what you have done), testing (what you have tested and how), data patterns (your results), and possibly either comparisons or interpretations of those results, depending on the purpose of the report. The conclusion will summarize your progress. Since the work is ongoing, the first four components are less likely to be included. A typical progress report will contain the following elements:

- Progress since the previous report;
- Problems encountered;
- Plans for the next stage.

> **A progress report claims that you have made progress toward the goal, supported by evidence.**

Each situation is different, so it is difficult to give specific guidelines on how to write, particularly since much of the variation depends on how much detail the boss wants. The only way to know that (other than trial and error) is to ask your colleagues. The most important idea to keep in mind is that a progress report is not just a list of what you have done, but a claim that you have made progress toward the goal, supported by evidence.

14.3 Technical reports

Engineers frequently write reports to propose a design solution, predict potential problems, or analyze actual design failures. These are *submitted to a supervisor with equal or greater experience in the same field who is not directly involved with the project but has background to understand it.*

> **Technical reports analyze or propose a solution to a single problem, and are complete without reference to other documents.**

Technical reports are the conclusion of a process of evaluation. Unlike progress reports, they should be complete in themselves without reference to other documents. Like progress reports, the focus is on a single issue. The importance and need are normally assumed—the reason the study and report were assigned is known to both parties. A technical report may include:

- Summary;
- Problem statement or goal;
- Evaluation and support (section titles specific to the problem);
- Conclusion or recommendation.

The report will often begin with a summary of the goal, the findings, and the conclusion. This introductory summary saves the reader time. The goal may be expanded into a problem statement that lists the criteria, constraints, or assumptions used in evaluating alternatives.

The majority of the report will be sections supporting the final conclusion or chief recommendations. This may include a theoretical model, simulations, or testing data, but citations are rare, on the assumption that the report recipient is familiar with background knowledge. Support may include evaluation of alternative designs or explanations, and whatever logic is appropriate to the situation: inductive, deductive, or abductive. More time will be given to alternative designs or explanations with more complex or important issues.

14.4 Management or consulting reports

In contrast to technical reports, management or consulting reports are *written for an audience who may not understand the technical details.* Management reports are internal company documents, whereas a consulting report is for a different organization. These reports will be *written differently depending on the goal and recipient.* For example, you might need to write different reports for the accounting and marketing departments, or for top management. Nevertheless, you are supporting the *implicit claim that you have done your work well and are presenting a fair assessment of the situation that will be beneficial to the one receiving it.*

> **Since the recipient may not read or understand all the technical details, an initial summary and organizational markers are crucial.**

In these reports, like technical reports, there is often an initial summary of the goal, key findings, and conclusions or recommendations. Since the recipient may not read the entire report, it is more important to use organizational markers such as introductory section summaries and phrases pointing to information found in other sections.

Fewer of the common components will be present, and less technical detail. Thus details of the framework and testing method may be reduced or eliminated. Instead of comparing various alternatives, more time is likely to be spent justifying the recommended course of action. This of course depends on what the recipient wants. Again the report should be self-contained, with few citations apart from industry standards or similar documents.

14.5 SWOT analysis

SWOT (Strength, Weakness, Opportunities, Threats) analysis is a popular strategy for organizing claims. It considers Strengths and Weaknesses within the organization and Opportunities and Threats outside it, relative to a goal (Figure 14.1). *Each box is filled in with claims appropriate to the category.* Support may be listed in the box as sub-points or placed in an accompanying explanation, where the claims of each category are more obvious (e.g., "One of our greatest strengths is . . .").

> In SWOT analysis, each box lists one type of claim.

SWOT analysis can be used in academia or industry for many purposes, from school accreditation to new product evaluation. *Separation of different categories of claims makes assessment easier.*

As with other reports in this chapter, importance and need are often assumed, or written separately from the SWOT analysis. The goal must be stated, as each category is related to that goal. Beyond that, the expected components are totally different from those used in academic writing, being the four categories on the chart. This illustrates that although the common components are useful for academic writing, other categories may be useful for other genres.

For example, some organizations develop their own progress report charts, following a similar pattern, but focusing only on internal factors. The example shown in Figure 14.2 is modified from a form used by the technical department of a local industry. It ensures that all areas are addressed in each report.

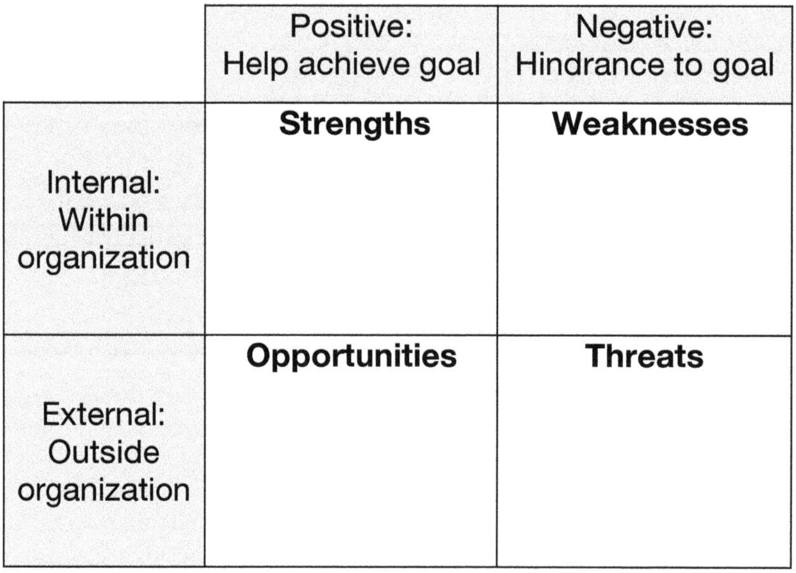

Figure 14.1 SWOT analysis chart.

	Positive	Negative
Current	Accomplishments since last update	Open issues and concerns
Future	90-day plan	Issues requiring help and support

Figure 14.2 Categorical progress report.

14.6 Summary

Each business or industry uses different types of reports and different reporting formats, but an attempt to generalize expectations is presented in Table 14.1. Just as once you have learned to write research articles you can modify that to any other form of academic writing, once you have learned to write one type of report it is easier to learn a second. The main thing to keep in mind is that the reports vary according to the goal, audience, and criteria, all of which affect the claims made and the type and amount of support needed.

Many other types of reports could be mentioned. White papers, policy papers, environmental impact statements, or safety analysis may be written for government organizations. Product specifications are written for business customers, and users' manuals for retail customers. Space limitations preclude discussing each in detail.

By analyzing exemplars, often available from coworkers if you ask, you will be able to write anything. Just remember that some genres will require other components, so add whatever is appropriate, and that not every report will be a good exemplar. Evaluate the samples you get to see whether the organization, claims, and support are clear. If not, include the same things, but try to stand out for the quality of your argument.

Evaluate your understanding

(1) What makes learning technical writing more challenging than academic writing?
(2) What accounts for differences between the reporting formats?

Table 14.1 Components and support in technical writing

Component/support	*Progress reports*	*Technical reports*	*Management reports*	*SWOT analysis*
1 Importance	No	No	Maybe	No
2 Need	No	No	No	No
3 Research goal	Maybe	Goal	Goal	Goal
4 Framework	No	Maybe	No	No
5 Research details	Yes	No	Yes	No
6 Testing methods	Yes	Yes	Maybe	No
7 Data patterns	Yes	Yes	Yes	No
8 Comparisons	Maybe	Maybe	Maybe	No
9 Interpretations	Maybe	Maybe	Yes	No
10 Conclusion	Progress	Conclusion	Conclusion	Conclusion
I Past research	No	No	No	No
II Present research	Yes	Yes	Yes	Yes
III Logical	No	Yes	No	Yes
IV Evaluative	No	Yes	Maybe	Yes
V Organization	No	Maybe	Yes	SWOT
A Timeframe	Maybe	Maybe	Maybe	Maybe
B Accounting	Maybe	Maybe	Maybe	Maybe
C Strengths	-	-	-	Yes
D Weaknesses	-	-	-	Yes
E Opportunities	-	-	-	Yes
F Threats	-	-	-	Yes

Homework

(1) Obtain exemplars for whatever you plan to write, if possible. Evaluate their argument structure and components before attempting something similar yourself.

(2) Talk with an experienced coworker to find out what your boss expects.

Teaching tip: Class assignments to write these genres lack intertextuality and therefore are not authentic. Nevertheless, if you have access to authentic texts, you can help students evaluate the quality of the organization and argument structure.

Chapter 15

Writing about yourself and others

15.1 Blowing your own horn

The idiom "blowing your own horn" often implies pride, and boastful confidence in your own ability, but I am using it in the sense of announcing your arrival, as horns used to be blown to announce the presence of an important person, or a returning voyager would sound his horn to announce his arrival. All the genres in this chapter have to do with getting a new position, for you or someone else. Thus your *basic claim is that you (or they) are the best person for the position.*

For any position, whether in graduate school, academia, or industry, there may be many qualified candidates, all with good grades and at least some experience. How will you stand out, getting to the shortlist, the interview, and finally the position? In order to claim that you are the best person for the position, you will need to do some research to find out what they are looking for, then arrange your support to show that you are that person. You need to not only be qualified, but convince them you are qualified. That involves writing a well-supported argument.

> The person who gets the position is not necessarily the most qualified, but the one who makes the best argument showing they are qualified.

15.2 Entrance essays

Two different types of essays may be important for getting into graduate school: those on standardized tests and entrance exams or essays administered by the school. Standardized tests include language proficiency tests and aptitude tests. A quick look at the rubrics for grading the latter reveal that they look at a student's ability to write an argument. Since there are many specialized test-preparation programs, I will simply remind you that you need to make clear claims and support them well with evidence and reasoning.

In addition to standardized tests, many schools require students to write one or more essays, either in the application or in person at the school. The former allows you time to write and rewrite, but is subject to excessive help from other individuals, while the latter

Find out all you can about the program and professors, so you can ask specific questions and indicate specific interest.

tests your ability to write quickly under pressure, with no assistance. If you prepare for standardized tests, it makes sense to prepare for these exams as well.

One of the main questions a school will ask is why you want to go there rather than elsewhere. Accordingly, your *main claim is that there is a good match between your interests and the strengths of the program.* This requires that you do some research, looking at what the professors teach, and what they research. Just like a grant application, it may require creative thinking to connect something you have done with their interest. For example, when my wife saw a poster offering a scholarship to study Austronesian linguistics, she had never considered that specialty. Nevertheless, she told the professor that as a child she had performed Austronesian dance. She got the scholarship and now has researched Austronesian linguistics for over 30 years.

As with any other argument, *using specific evidence to support your claim is essential.* If you can write coherently about the concepts of a field and show why you could plausibly be interested in the specialty of one or more professors, based on something you have done before, you will have an edge.

15.3 Job applications

Job applications involve many elements. There may be a standard form for basic information. They may also require a cover letter and resume. In general, you should not duplicate information, but can highlight things written in another portion.

Like entrance essays for graduate work, make sure you do your research, not only about the job itself but also the company. This shows your interest and allows you to talk knowledgeably in an interview, but also allows you to *adjust your claims and slant your evidence in the resume to highlight items of interest to the employer.*

Writing is important in any job, so write well on the application and list any publications or presentations.

Writing is an important part of any position in science or engineering. You have learned how to write, and hopefully write well. Show that skill. Work on publishing early and often. Even undergraduate work can be published in some journals. If you worked with a professor and were second author, that is still far better than nothing. Presentation at conferences shows your experience with both public speaking and presentation software. If the conference allows it, write a conference paper.

Other things count too. What is unique about you? If you have skill in graphic design (illustrations), or finance (grant proposals), that is important. If you spent time in a different country, learning a different language and culture, you have learned to think and communicate in creative ways. Starting your own business, even if it did not make money, shows initiative. Lots of applicants have good grades and recommendations. What do you have that they don't? *To support your claim that you are the best person for the position, you must provide specific evidence.*

15.4 Resumes and CVs

The terms resume and CV (curriculum vitae) are often used interchangeably, but there are differences in length and purpose. Most job applications require a resume. A resume is short, one or at most two pages, highlighting your most important qualifications. A CV is a complete list of your accomplishments, and is usually reserved for research positions, grant applications, or personal information on your website.

Data never speak for themselves. You are trying to get a potential employer to see your record in a certain way—that you are highly suitable for the job. You are also trying to draw attention to anything that will make you stand out from the crowd. Thus layout is far more important in a resume than a CV.

There are two approaches to resume writing. One suggests a simple one-page resume that can be used with any position, with additional information placed in the cover letter. The second suggests a longer two-page resume tailored to each job. Arguments could be made for each. The first is simpler, but cover letters may not be read carefully.

> **Keep an ongoing CV that lists all your accomplishments, then create your resume from that.**

No matter which method you choose, I suggest that you create a CV that you update regularly (at least once a year) with your jobs, publications, and other information. Just like "write as you go" for the details of your research, it is easier to add the details early than to search your files to find them later. Your CV should include categories for:

- Name;
- Contact information;
- Education and training;
- Degrees, licenses, certifications with dates and granting agency;
- Job experience, both employment and self-employment;
- Specific job accomplishments including major reports;
- Publications and presentations;
- Awards, honors, and organizational affiliations;
- Reference contact information.

You may add categories such as military service, security clearances, travel, volunteer experience, foreign languages, or others relevant to your field. Adding upcoming papers or conferences will remind you to go back later to fill in the details.

When it comes time to write your resume, the first five elements listed for the CV should be included. Others are optional. You may choose not to list every detail, particularly short-term or part-time jobs, but beware of gaps in your school and employment history.

Organizing your resume is an art. While many books and websites give guidelines, remember that you are trying to *highlight the strongest evidence that you are the best candidate for the job*. Whatever that is—employment, education, or licensure—should be placed front and center where it cannot be missed. That will change as you gain experience, so your resume should change as well.

Material that will not fit in the resume can be added to your job application as supporting information. The employer will not look at this unless they are impressed with your resume, so make sure that whatever makes you stand out is on the resume.

15.5 Reference letters

If you become either a professor or a manager, at some point you will be asked to write a recommendation letter for a student or employee. If they have done exceptionally well, this is easy. If they have not done well, you probably want them to move on, but have an obligation to be honest. Either way, *your goal is to help them get the position, so the implicit claim is that they are the best person for that position.*

Potential employers read between the lines of a reference letter, looking for what is not being said as well as what is being said. Thus recommenders may try to write between the lines so the message is clear, while avoiding direct criticism. Any direct mention of a negative characteristic will kill a person's chances, unless it is the flip side of something extremely positive. For example, saying someone is extremely diligent in the lab but lacks confidence in front of people alerts the potential employer both that they may not show their best face in an interview and that they may not be suitable for a position requiring high-profile presentations.

Reference letters usually have five parts:

- Self: why were you chosen to write this reference?
- Situation: how long have you known the person and in what relationship?
- Strengths: what are their greatest strengths (and weaknesses)?
- Suitability: how well suited are they to the position?
- Summary: how strongly do you recommend them?

First it is important to introduce yourself, and state why you are qualified to give a recommendation. Next you need to specify the situation in which you have known the person, and for how long. Often these are combined in one paragraph.

> Specific incidents and skills show whether someone is suitable for a position.

Next comment on the greatest strengths you have observed directly in your contact with the individual. Specific incidents or supporting evidence are helpful. In order to comment about suitability, you need to know the job requirements, so make sure you get those from the individual when you agree to write the recommendation. A person may be very good, with many strengths, but not be suitable for a particular job. Honesty requires you to state or at least imply your reservations.

The final paragraph is a summary, including a statement of how strongly you recommend them for a position. Without reservation is the strongest compliment and should be reserved for exceptional individuals.

Reference letters for entry-level positions are usually less than one page. Those for higher-level positions may be two pages. Anything beyond that should be added as supplemental material.

15.6 Summary

Good writing is always an asset in getting a position, but not the only factor. Luck and relationships play a role as well. One time while waiting for interview for a graduate assistantship, I was reading a book on string theory. The professor saw it and commented he had just finished reading the same book. That gave us a good starting point that was not related to the position itself. Connections of any kind give an edge.

The genres discussed in this chapter are very different from other academic writing. Thus it should not surprise you that they contain different components. Nonetheless, there are similarities in structure. Some sort of introduction says why the reader should pay attention, followed by the goal and reason for pursuing it. Supporting evidence must be specific and support the claim that you are suited to the position. A conclusion summarizes the argument.

Evaluate your understanding

(1) How are the genres discussed here like other academic writing, although the components included differ?

Homework

(1) Begin to build your own CV so it is ready when you need it.

Chapter 16

Academic and technical presentations

16.1 You are the expert

Many novice engineers and scientists are quite comfortable working alone in a lab, but far fewer exude confidence when they step in front of a group to present their research. Nevertheless, just like writing, the more you do it, the better you will get and the more confident you will feel.

> Remember that no one knows more about your topic than you and everyone wants you to succeed.

Remember that you are the expert. No one knows your work better than you do, so prepare well and speak confidently. Everyone in the audience has been or will be in your position at some point and, unless you have made a lot of enemies or are in a competitive job interview, wants to see you do a good job.

Many written documents are also presented orally. After examining some basic principles we will consider three genres: academic oral and poster presentations and technical presentations.

16.2 Different format, same principles

Although the format of oral presentations is different from written work, similar principles apply. Just as you examined exemplar articles before starting to write your own, analyze similar presentations, then prepare in stages. As with writing, this will help you fit the expectations, yet stand out for excellence.

16.2.1 Preparing in stages

As you listen to other presentations, don't be a passive observer. Take notes, not only on the content but on the structure. How did they begin? Which components did they include and how many slides were devoted to each?[1] Was the total number of slides suitable for the time, or did it seem too fast or too slow? How did they conclude? Think about which talks caught and kept your attention. Was it the topic that interested you, or was it a good presentation? Comparing notes with a friend can help distinguish the two.

Because of time constraints, some components important in written work may be reduced or eliminated in a presentation. For example, there may be less emphasis on showing the research gap or need for the work. Standard methodologies or tests may be mentioned without details. In engineering you may explain only the most novel design details. Your research goal is essential, as is evidence showing you achieved that goal and made a contribution.

Once you know what you need to include, *start preparing early*. I suggest following the stages shown in Figure 16.1. Stage 1 is the equivalent of prewriting and writing. Stages 2, 3, and 6 correspond to the three major phases of rewriting, while 4 and 5 address special requirements of a presentation. As with writing, allow at least half the time for several rounds of revision, and additional time to practice.

> **Just like writing, prepare a presentation in stages, beginning with the large scale and moving to smaller changes.**

In stage 1, think about argument structure. What claims do you want to make and how will you support them? List the components you plan to include. Begin with titles and a few keywords on each slide. Each slide should address one main point, just like a paragraph. Slides are free, so use them generously, without multiple layers or squeezing too much on one slide. On the other hand, avoid large blank spaces, which the audience will subconsciously interpret as gaps or weaknesses in the argument.

Next, organize the content, changing the order of slides and points on each slide so the audience can follow your argument. Introduce vocabulary and abbreviations as needed, rather than at the beginning, since the audience cannot look back as they can in a paper.

During stage 3, flesh out the slides, but keep the number of words to a minimum. Use keywords or bullet points rather than sentences or paragraphs. Too many words on a page will tempt you to read rather than speak. More importantly, it will distract the audience, as they will not know whether to read or listen. Use graphics liberally, but if several are similar, choose one as an example rather than showing all.

Once the basic order is established, think about how to attract interest. What is unique about your work? Why should other people care about it? How could they use it or build on it? What social relevance does it have? Is there any controversy related to it? Any unexpected findings? Add an introduction that will draw people in and a strong conclusion.

1. Lay out the content—titles and keywords for each slide
2. Organize the content—cut, add, and reorganize slides
3. Clarify the content—add visuals and text to each slide
4. Attract attention—strong introduction, conclusion
5. Maintain attention—animations and transitions
6. Finalize the content—proofreading and revision

Figure 16.1 Stages in preparing a presentation.

At this point the basic content of the talk should be ready, but you are not done. Add animations to direct people's attention to each item in the order you will talk about them.

Finally, go through the whole presentation, checking for spelling or formatting errors. Then set it aside for several days or a week. When you look at it again, you will undoubtedly notice irrelevant or unclear portions and important points you missed. Fix them. Then practice it with your colleagues, if possible, and modify it again based on their feedback.

16.2.2 Tell a good story

> You do not have time to present your full argument, but everyone loves a good story. Make yours believable and compelling, and interesting to your audience.

Just as a research article is not merely a report but an argument, so a presentation is not merely a report but a story. Presentations are too short for a full argument, but must present a compelling and believable story. People get bored listening to reports, but never tire of a good story.

First you must catch your audience's attention. As one author writes, "During the first few minutes of your presentation, your job is to assure the audience members that you are not going to waste their time and attention."[2] This begins the moment you step on stage. If you appear and sound confident, even if you are shaking in your shoes, the audience will be alert. Don't lose their interest. Just like a research article, first highlight the importance of the work, leaving the outline for the end of the introduction.

The more mundane your results, the more crucial it is to present them in an interesting way. The best presenters are not necessarily those with the best research, but those who tell an exciting story. Fair or not, they are the ones that the audience will remember.

16.2.3 Focus

There is a symbiotic relationship between audience and presenter. If you are excited about your work, let that show. Excitement is contagious. If the audience is looking at the clock wondering when the agony will end, you will feel the same. Distraction is deadly, so do your best to eliminate it.

> To maintain audience attention, the number of slides, overall design, page setup, fonts, positioning, and amount of information are all important.

The first distraction is the number of slides. With 10 slides in a 20-minute talk, each will be on the screen for two long minutes. Unless you are an excellent orator, the talk will also seem to drag. On the other hand, trying to fit 60 slides into 20 minutes allows only 20 seconds per slide, leaving the audience overwhelmed and you out of time. Depending on your speaking style and the amount of information on each slide, one or two slides per minute is about right for academic presentations. Fewer slides may be better in business or industry if details are important.

A second distraction is slide design.[3] Academic and technical presentations use simple background designs to avoid cluttering the screen and restricting the space. Moving backgrounds or images are particularly distracting because our eyes and minds are naturally drawn to motion. Also, make sure the page setup (on the design tab) matches the screen you will use. The default page setup may be "widescreen (16:9)" to fill the *computer* screen, but most classroom *projection* screens are "standard (4:3)."[4] Widescreen format will cause the font and graphics to appear smaller. Changing the page setup alters the proportion of images and text alignment, as shown in Figure 16.2. Changing the design has a similar effect.

Make sure your slides can be read from the back of the room—at least font size 16, even on graph axes, preferably 24 or larger. It is easier to read sentences written in serif fonts (with small lines at the end of strokes, like Times New Roman and Cambria), while san serif fonts (without those lines, like Arial and Calibri) are more suitable for headings or small letters. Calligraphy, Comic Sans, and other fancy fonts are generally best left for non-academic uses.

Another source of distraction is having too much on one slide. A presentation should be spoken, not read. Spoken English incorporates repetition, pauses, and fillers to give listeners time to process the message. The material on the slide should supplement the talk, not distract from it. Alignment of text on a slide is also important. Figure 16.3, explaining why it is better to introduce abbreviations as needed, demonstrates how placing related text on one line allows our brain to process it more easily and thus retain it better.

Complex graphics can also be distracting. Rather than reproducing graphics directly from your writing, redraw them to highlight what you want your audience to focus on. Limit complexity, using color, borders, or animations to draw attention to important features. Frames or donuts mark items faster and more effectively than a laser pointer, and remain on the screen.

Remember that the resolution and color of the computer projector is not as good as your computer screen. Whenever possible, crop pictures to show only the area of interest, then resize them, maintaining the proper aspect ratio. Correct the brightness and contrast to make them easier to see. Add arrows to point to important features, with or without a label. Positioning text over a portion of a figure that is hard to crop out allows you to increase the size of the figure while simultaneously covering the distraction.

Animations can effectively focus attention, either by sequentially revealing lines of text or highlighting with frames or arrows, but if overused or too dramatic they can also be distracting. Setting animations to appear after the previous one, with an appropriate delay, can reduce mouse clicks and keep you on pace.

> Animations and transitions can focus attention if they help add meaning, or can be distracting.

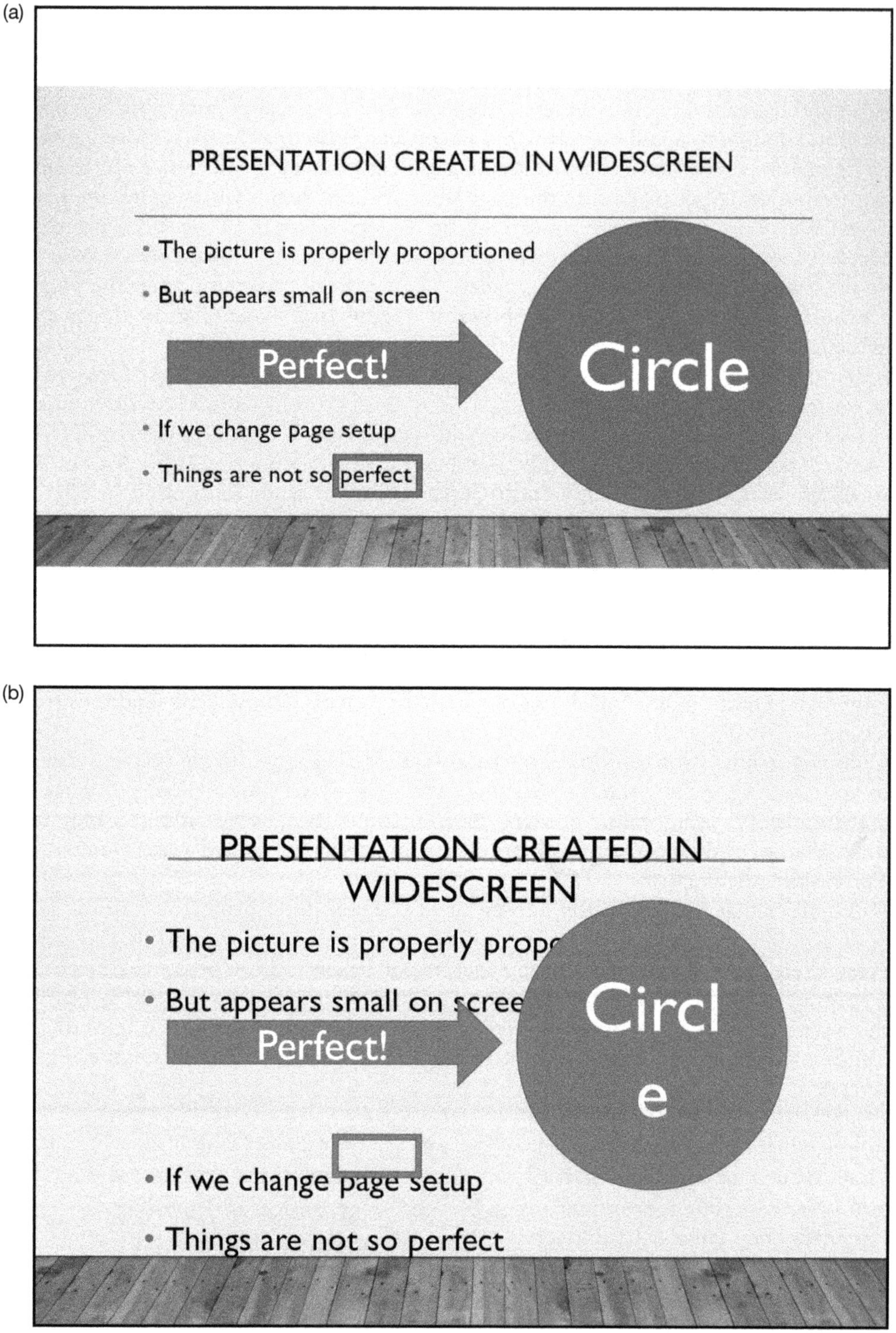

Figure 16.2 Comparison of projected images (a) as laid out in 16:9, and (b) changed to 4:3.

(a)

(b)

Figure 16.3 Alignment of text can (a) impede or (b) facilitate understanding.

Similarly, transitions should add meaning rather than distracting.[5] Fade connects slides, while a turning or changing motion can indicate shifts between sections of a presentation. Very dramatic transitions have a place in social or motivational slide shows, but not academic presentations.[6] Make sure you go through the whole presentation in slideshow mode, preferably on the computer and projector you will use, to check everything before presenting.

16.2.4 Finish strong

Feeling comfortable during the question and answer time will help you finish strong. Perhaps the most important thing is anticipating questions that may be asked. There is nothing more empowering than hearing a questioner ask a question, and feeling, "Yeah, I got this!" Practicing with your lab mates can help you both predict the questions and practice the answers.

You can even drop hints to get people started in a direction. For example, during my doctoral defense on tomato breeding, one of the examiners was a plant physiologist, so I purposely mentioned that botanically a tomato is a berry, expecting him to pick up on the clue and ask about it, even though it had no direct bearing on my research. He did. As soon as he started down that track, I knew exactly what the series of questions would be and had planned how to answer each, and had to suppress a smile.

If you spend less time looking for a slide, you can focus your attention on answering the question and have more time to do so.

Make it easier for the audience to ask questions by adding slide numbers. In slideshow mode, typing a number followed by "Enter" causes that slide to appear. If the questioner does not remember the number, type "Control-S" (Show)[7] to bring up a box listing the slide titles, then click the appropriate one. Both tricks save time and allow you to focus on formulating your answer (and are impressive to those who don't know how to do it).

You can also hide a slide you might need during the Q&A time. For example, you might show one graph. Similar graphs can be marked with "Hide slide" and will not show until you move there by slide number or title.

Many of these tricks, and others, I have learned from watching presentations. If you see something that is interesting, but do not know how to do it, ask! The presenter will undoubtedly be happy to tell you.

16.3 Academic oral presentations

Class presentations, oral thesis defense, conference papers, and job interviews have different goals and audiences, thus emphasize different things.

There are at least four different types of academic oral presentations: class presentations, oral thesis/dissertation defense, conference papers, and academic job interviews. As with writing, it is important to consider the goal, audiences, and criteria for success for each.

For a class presentation, normally you must show you have mastered the class content. Class presentations have two audiences: the professor and the rest of the class. While the class wants an interesting presentation, the instructor is looking for certain information or demonstration of skills specified in the assignment. Since your immediate goal is to pass the class, obviously the professor is the primary audience, but since the class is the majority, don't forget their needs.

Graduate degrees and some honors programs require a concluding oral presentation or defense of your thesis. Since by that time you have completed both your research and writing, and know everyone in the audience, it should be a celebratory event. Nonetheless, because of the high stakes, it is often feared. Remember that the audience is on your side. They want you to succeed. Your advisor is convinced you are ready. If you can clearly present what you have done and how it contributes to the field, you will get your degree.

Conference presentations not only present your research findings, but also showcase you as a researcher. While your immediate goal may be to fulfill a graduation requirement, others are evaluating both your research and presentation ability. Unless there is a prize for the best presentation, the criteria may not be stated, but novelty of your research and clarity and interest of the presentation are always important. As with a research article, remember that your audience includes both experienced researchers who want to know what you have added to the field and neophytes who need a broader picture.

When planning an academic job interview, focus on your immediate goal: getting the job. You may be most comfortable talking about the main findings of your research, but if the potential employer is interested in something else, try to identify some aspect of your research that more closely matches the job description. In this case, your main audience is those who have a say in the decision, which normally does not include graduate students, so your talk should include more technical detail, particularly any theoretical aspects of the work.

16.4 Academic poster presentations

In addition to oral papers, many academic conferences have poster sessions. There are many differences between the two, including the format, the time you have for presenting your work, and the number of times you may do so.

First, let me explain how a poster session works. Your poster, usually no larger than 36 × 48 inches (90 × 120 cm), is mounted on a wall or movable divider. During the poster session participants wander by, look at your poster, and ask questions if they are interested. Thus it is important to have an eye-catching poster and be able to give a quick but interesting summary of your research, so they will be willing to stay and listen to more details.

Unlike an oral presentation, you cannot plan to speak in a fixed order but must be able to interact with the questions posed. If you get a general question to talk about your work, summarize the main contribution within two minutes. While speaking, be aware of the body language and cues given by the listener. If they are interested, continue. If they keep glancing away and do not ask follow-up questions, ask if they have any further questions, and if not thank them for their time and let them move on. As people come and go you may talk about your research several times, focusing on a different aspect each time depending on their interest.

A poster makes the same argument as you would in an oral presentation on the same topic, but because of the format the organization is not as straightforward. *Make sure that the main contribution of your research is highlighted and central*, not tucked away in the bottom corner. Visuals, not words, attract attention. Leave the details and longer explanations as answers to questions, rather than trying to squeeze everything onto the poster.

A good poster can be read sequentially, combining graphics and text in a way that your eyes are naturally drawn to the starting point and from there proceed naturally in the order intended by the author, which may not be linear. The graphics by themselves should tell much of the story, supplemented by details in the text. There are many ways to organize a poster to accomplish this. Figure 16.4 shows some alternatives.

A poster need not be linear, but must have a logical structure, drawing attention to the main points.

Figure 16.4a shows a design that might be useful in chemical engineering to describe three steps of fabrication of a new compound, followed by characterization to demonstrate

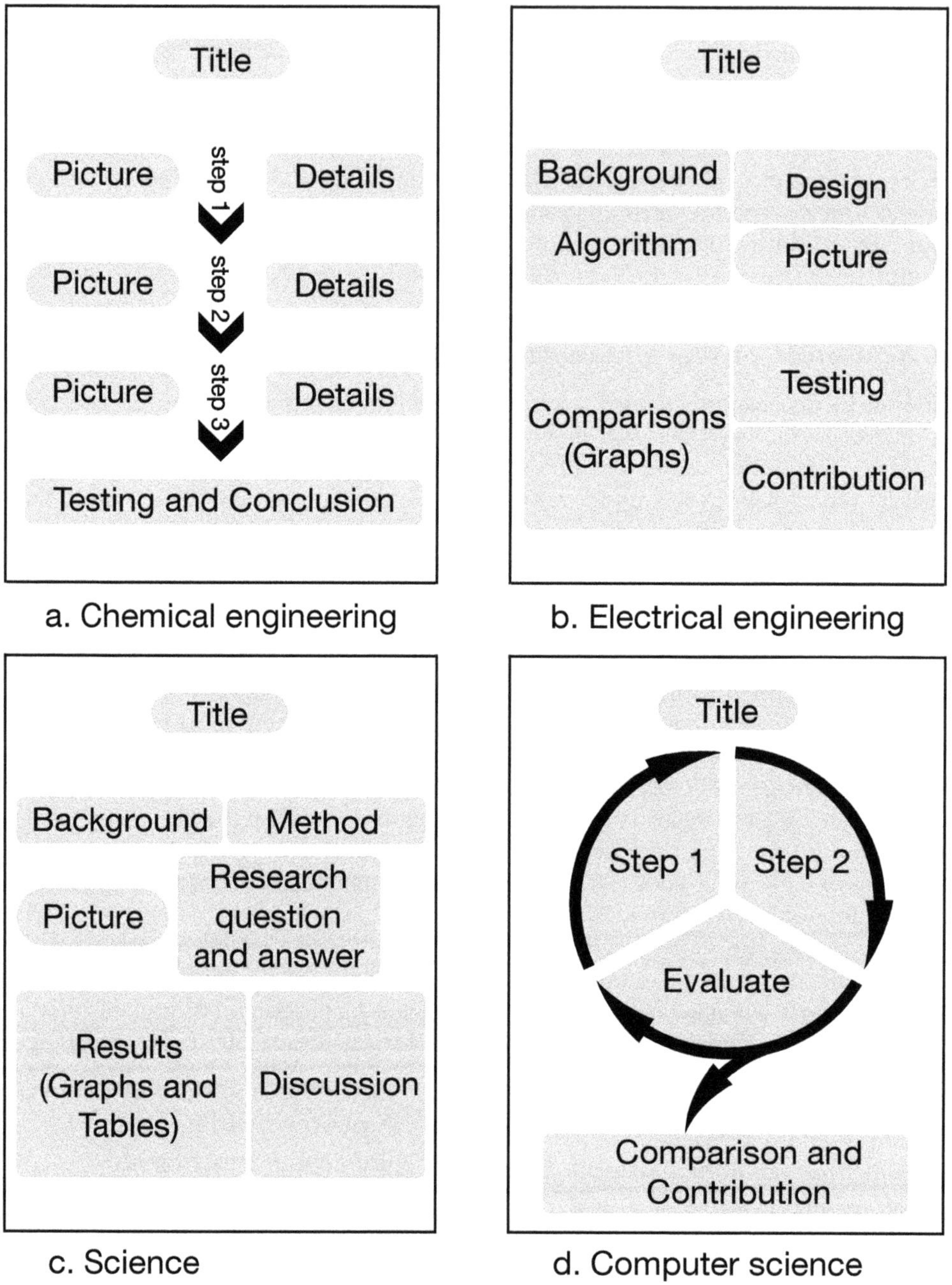

Figure 16.4 Possible poster designs.

its formation. For each step, a picture and brief description are placed on each side of an arrow with the name of the process.

Figure 16.4b illustrates a format that could be used in electrical engineering to present a new design. The top half describes and illustrates the new design, while the bottom describes the testing and comparison with other designs.

Figure 16.4c shows how an IMRD format paper could be laid out in four unequal quadrants, with the unifying research question and answers in the middle. The picture helps draw attention to the middle of the poster, and thus to the questions.

Figure 16.4d illustrates how a cyclical algorithm or other cycle could be shown, with each step in a section. Comparison with other algorithms and the benefit of the research are summarized at the bottom.

Although it is hard to show in such small illustrations, space is important in a poster. It should look full, but not cluttered. More important elements should be in larger font, so the viewer's eyes are drawn to them. Graphics should be closely connected with the accompanying text, separated from other blocks of graphics and text so the connection is clear. Blank space can be as effective as a line in directing the viewer's eyes.

You are attempting to give a quick visual representation, not all the details. If you are considering landscape (horizontal) orientation, check the conference regulations, as some only allow vertical. Sketch different possible layouts before creating a rough draft using PowerPoint or a specialized program. After filling in the text you will have to make adjustments. Use bullet points rather than sentences and paragraphs. Make sure you reserve space for the title, author, and your department and school. Just like writing, after you are clear what needs to be included, it is sometimes better to set the whole thing aside and try again.

16.5 Technical presentations

There are as many types of technical presentations as there are technical reports. The technical presentation accompanying a position paper, project proposal, progress report, or concluding report will differ, because of the different goals. Audience is also important, the main question being the background and interests of your target audience.

Many of the principles for academic oral presentations also apply to technical presentations in business or industry: timing and number of slides, simplifying graphics to show the main point, highlighting information. Begin by laying out your argument, based on the format of the related technical report. Fill in the details and revise to clarify. Catching and maintaining interest is less important because of the inherent interest of the parties involved.

Nevertheless, since technical presentations in business or industry involve many considerations different from academia, I encourage you to consult other resources for more details. From a single technical report you might need to prepare four different presentations for managers from research and development, production and quality control, sales and marketing, or accounting and finance.

> **Your presentation must provide the information needed by your audience.**

Evaluate your understanding

(1) Why is it important to begin to prepare your presentation early?
(2) Why should an oral presentation be more like a story than a complete argument?
(3) How can you keep the attention of the audience?
(4) What techniques will help you feel confident while answering questions?
(5) How does presenting a poster differ from an oral presentation?
(6) How do technical presentations differ from academic presentations?

Homework

(1) Prepare a presentation, following the principles in this chapter.

Hint: Practice presenting it to at least one other person and get their feedback on how to improve it before presenting it publicly.

Teaching tip: Try to provide good examples throughout the course in your own teaching, but also learn from techniques your students use.

Teaching tip: Many books specialize in oral presentations, so this chapter is not comprehensive, but shows how to apply the principles learned and provides practical hints from personal experience.

Notes

1 Take a copy of the component list and note how many slides discuss each. Add other components such as timeline or budget if appropriate.
2 Dale Ludwig and Greg Owen-Boger, *The Orderly Conversation*, quoted at http://thepitcher.org/10-quotes-on-the-importance-of-mastering-public-speaking.
3 I will be using terms from PowerPoint, the most common presentation software. The same principles apply to other programs, although the terms may differ.
4 Large lecture halls may use widescreen format. Ask if you are not sure.
5 In linguistic terms, animations and transitions should be semiotic, with each type used to convey a meaning.
6 For this reason I do not recommend presentation programs like Prezi for novice presenters. While I have seen some excellent talks done with it which use the ability to zoom in or out or return to a previous section to highlight the storyline, the movement can be extremely distracting if not semiotic.
7 Command-S on an Apple computer.

Chapter 17

Principles for successful email

17.1 Importance of email

The final genre we will consider is email. Email is currently the most common medium of communication in academia, used for everything from departmental memos and teacher–student interaction to corresponding with colleagues or editors in other countries, making email literacy an essential part of academic English.[1] Other electronic social media are rising in importance, especially among young scholars, but whether they will supplement or supplant email remains to be seen.

Writing email might initially seem to have little to do with other academic writing, but there are many similarities. Without explicit instruction you could gradually pick up the skill, just as you would eventually learn the expectations of academic writing, but might fail many times in the process. Like other writing, important emails must be worded just right, so first thinking about the argument structure, then writing and rewriting in stages, can lead to more successful email.

> Applying the principles of argument structure and rewriting in stages can improve your email communication, especially for formal email.

Since email involves so many different goals, situations, and relationships between sender and recipient, it could easily be broken into a dozen genres, but that awaits further study. For now we will focus on principles that will help you achieve an appropriate level of formality in any situation.

17.2 General email etiquette

First let me propose a few general principles of email etiquette, listed in Figure 17.1. These hold true no matter the purpose or recipient of your email.

One of the advantages of email is that it provides almost instantaneous correspondence anywhere in the world. The related disadvantage is that we have grown accustomed to an almost instantaneous response. If you need more time to think or act, send a quick acknowledgment and say when you will be able to provide a more complete answer.

Always use a subject line that summarizes the topic of the email, both to indicate that it is a valid email and to make it easier for the recipient to find a week later. Similarly, make sure you include at least one personally identifiable line of text in the email, especially if you include an attachment or web link, so the recipient knows it is from you and safe to open.

1. Respond promptly;
2. Use the subject line and identify yourself;
3. Send it to the right person;
4. Be careful what you say.

Figure 17.1 General principles of email etiquette.

> Principles for email are no different than any other communication. Consider the situation and respond appropriately and politely.

Before choosing "reply" or "reply all," think who needs to see the response, and before you hit "send," double-check the list of recipients. In a series of correspondence with a coauthor and editor of a book chapter, I sent an email to all and my coauthor replied, asking for a response from the editor. A week later I was wondering why we had not heard anything, then noticed that the reply had gone only to me, even though the body of the email was addressed to the editor. On the other hand, when working on a book on email, the editors were corresponding with an author, asking for some changes. The author sent a very polite response, with copies to the coauthors. Unfortunately, in the subsequent exchange the authors continued to use "reply all," forgetting that the editors were still receiving those emails, some of which were definitely not intended for our eyes! (We still published their chapter, after revision.)

This leads to the final principle: be careful what you say. Remember that an email is an electronic document, easily sent but impossible to contain thereafter. Always be discreet because you never know who might see it at some point, as people have found with images texted to friends or notes posted on what they thought was a private conversation.

17.3 Position and imposition

Before discussing factors that can improve your chances of success in email, we need to consider different styles of email, how formality affects style, and how position and imposition determine the formality required. The rest of this chapter will then be devoted to the question of how to write email that involves a high level of formality, such as letters of request, invitation, or apology.[2]

> Email can be used for purposes ranging from informal texting to very formal or face-threatening letters, leading to a range of styles.

Email is a *means* of communication, not a *genre* of communication. Just as ink on a page can be used for a quick note, a letter to a friend, or a formal business letter, email can be used for a range of genres from texting format to informal memos to formal letters.

The first task, then, is to evaluate how formal a style is necessary. The formality required depends on three factors: power, distance, and imposition, as depicted in Figure 17.2. Power and distance are measures of the relative position of the sender and recipient, while imposition is based on the contents of the email. Although there is some interaction between the factors, in general the formality required is determined by the highest arrow.

Power refers to the degree of power the recipient has over the sender, often based on their relative status. Letters from someone of low status to high status require greater formality than from high to low. For example, a graduate student writing to a professor or journal editor is expected to write in a more formal style than is required for the reverse. As with any communication, what is considered appropriate is largely determined by whoever holds power in the interaction.

Distance refers to how well the sender knows the recipient. The better you know the person, the less formality is required. Thus letters from a graduate student to his or her advisor may not need to be as formal as to another professor.

Imposition requires a little more explanation. Imposition in this case can be defined as demanding someone's attention or commitment, thus is closely related to the time or effort necessary to respond. Some emails merely convey information, requiring reading but no response, and thus involve minimal imposition. Others require a quick response, but no additional time commitment—a low level of imposition. Email that requires a longer, detailed response or additional thought or action carries a medium level of imposition. Many emails entail an extended commitment of time on the part of the recipient, and thus a high imposition, including requests, invitations, and many apologies, which often contain related requests. In these cases, different devices can be used to minimize the feeling of imposition, as will be discussed below.

Table 17.1 shows the relationship between formality and style, and examples of email content suitable to each level of formality. Success in email, especially in high-imposition situations, depends on an appropriate level of formality and use of techniques to reduce the feeling of imposition.

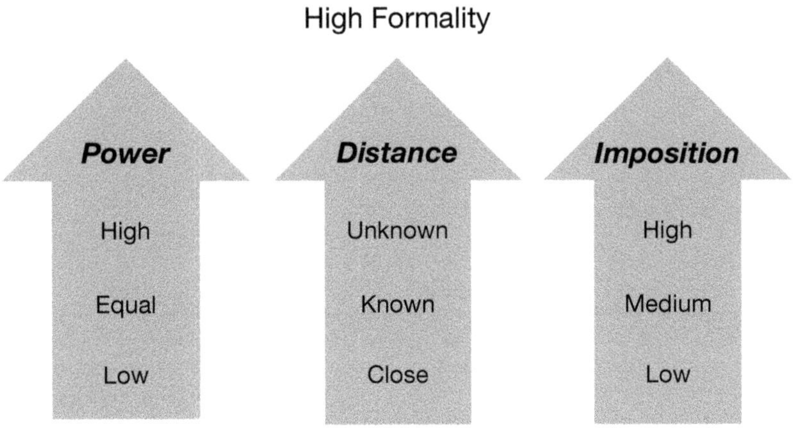

Figure 17.2 Formality required is determined by the tallest arrow.

Table 17.1 Formality and style in different types of email communication

Formality	*Style*	*Examples*
Formal	Formal salutation Body paragraphs Closing and signature	Request, invitation, apology Question—long response
Informal	Informal salutation Sentences, capitalization Informal closing	Ongoing discussion Informative memo
Very informal	No salutation Text language, emoticons No closing	Notes between friends Email chatting, texting

17.4 Five factors affecting email success

The rest of the chapter is devoted to factors affecting the success of email in situations that require a high level of formality, such as apologies, requests, or journal article submissions. As shown in Table 17.2, each is related both to a general communication principle and a writing principle described in this book.[3]

For informal email between friends, these factors may not be that important. Formal emails, on the other hand, like any formal writing, may need to be rewritten several times, paying attention to every word. It is often better to compose such letters offline in a text editor, then paste it into the email once you are satisfied. More than one email has been sent in an incomplete form because the sender hit the wrong key or clicked in the wrong place.

17.4.1 Terms of address

> The salutation and closing show relationship, how you fit in the academic community, which changes over time.

Terms of address, how you refer to the recipient in the salutation and to yourself in the closing, follow an expected format. This is not merely a formality, because these terms of address reflect relationship, an important element in any communication. Just as using the accepted

Table 17.2 Factors affecting email success or failure

Factors affecting success	*Principles*	
	Communication	*Writing*
1 Terms of address	Relationship	Fitting in
2 Justifying your request	Cooperation	Argument structure
3 Amount of information	Clarity	Coherence, conciseness
4 Reducing imposition	Politeness	Connection, connotation
5 Attention to detail	Intelligibility	Correctness

format for academic writing shows that you belong to an academic community, using the proper terms of address in academic communication shows *how* you fit into that community, which will change over time as you become better known and established as a researcher.

In formal academic correspondence, it is usually best to address someone based on the highest title or position they hold. Thus, although it would be appropriate to address someone as Mr. or Ms. [family name] in a non-academic context, in an academic context it is better to call someone Dr. [family name]. Since someone could hold a doctorate but not be a professor, Professor is a higher term of address. If that person holds a higher rank, for example Dean or President of an academic institution, they should be addressed by that rank. Salutations appropriate for various levels of formality are shown in Table 17.3. Professor may be abbreviated to Prof. if used with a family name, but not when used by itself.

For those using English as a lingua franca, remember that titles attach to the family name, not the given name. Furthermore, although the Chinese word for teacher is considered an appropriate term of address in Chinese, even for university professors, that is not the case in English. This leads to various terms of address being considered inappropriate, as shown in the last row of Table 17.3.

Context is vital. Even if you know someone personally, and in a different context might address them by first name, if you are writing a formal letter it is generally better to use the appropriate title, at least in the initial email. Subsequent emails in a series are often less formal, especially if exchanged rapidly, and may not contain either salutation or closing. Asian students may feel this is cold and impersonal, whereas a Western professor may be using it either as a sign of closeness or merely for efficiency, just as you might greet someone at the beginning of a conversation but not at the beginning of every exchange in the conversation.

Newcomers to the academic community often do not realize that not only do terms of address reflect relationship, they can also be used to negotiate relationship.

Table 17.3 Academic salutations suitable to different levels of formality

Formality	*Address*	*Title*	*Name*
Formal	Dear	President	Family name
		Dean	Family name
		Professor/Prof.	Family name
		Dr./Mr./Ms.	Family name
Informal	Hello	Professor/Prof.	Family name
	Hi	Professor	-
		-	Given name
Very informal	-	Professor	-
		-	Given name
		-	-
Inappropriate		Professor/Prof.	Given + Family name
		Professor/Prof.	Given name
		-	Family name
		Teacher	Family name
		Teacher	-

Closings and your signature can be used to negotiate a new relationship.

This is accomplished in the closing and signature lines at the end of the letter. Whereas at the beginning of the letter you are showing your respect for and distance from the other person, especially if presenting a request, invitation, or apology, by the end you are often trying to establish a closer relationship.

In Table 17.4 you can see a listing of various closings, from most to least formal.[4] "Best" is a new variant on "Best wishes" or "Best regards" and seems to be increasing in popularity as a closing for emails. "Thank you" was never considered an appropriate closing for paper letters but is also increasing in popularity. A closing common among Chinese is "Your student." Like the salutation "Teacher," this is a translation from an acceptable and common Chinese form, but not recommended in English.

Different types of signatures are illustrated in Table 17.5. In the signature line you are not merely identifying yourself but hinting at the relationship you would like to have with the recipient. Thus, a very formal signature indicates a desire to maintain a distance, at least in the present interaction. On the other hand, an informal signature indicates a desire to reduce the distance in subsequent emails.

An illustration of this sort of negotiation of relationship can be seen in this series of emails I exchanged with a professor, a well-known author in English for Academic

Table 17.4 Closings, from most to least formal

Closing		*Use*
	Most formal ↑	
Respectfully yours, Respectfully		Legal documents
Sincerely		General, emphasizing distance
Best wishes, Best regards, Best		General, emphasizing relationship
Cordially, Warmly, Warm regards		General, emphasizing close relationship
Thank you, Take care		Informal
[no closing]		Very informal
	↓ Least formal	

Table 17.5 Types of signatures, from most to least formal

Signature		*Example*
	Most formal ↑	
Given name, Family name, Degree		Gerald Rau, Ph.D.
Title, Given name, Family name		Professor Gerald Rau
Title, Family name		Professor Rau, Dr. Rau
Given name, Family name		Gerald Rau
Given name		Gerald
Nickname		Gerry
	↓ Least formal	

Purposes, who was to be a keynote speaker at a conference I would attend. I began by writing the following email.

> Dear Prof. [family name],
>
> I am writing a textbook, tentatively entitled *Writing Your First Science or Engineering Article in English*, . . . I was therefore very excited to find that you will be a keynote speaker at [conference], and would like to ask if it might be possible to meet and talk with you briefly while you are in [location].
>
> . . .
>
> Sincerely,
>
> Gerry Rau

I began by addressing the professor by title and family name, and closed with a formal but distant "Sincerely," but in the signature line I did not use a title, indicating a desire to move to a closer relationship. By using my nickname with my family name (my formal given name was visible in the standard signature panel at the bottom of the email), I hoped to reinforce my intent.

The reply was as follows.

> Dear Gerry, if I may,
>
> I would be happy to meet with you while I am at the conference. Your work sounds very interesting and I would like to learn more. . . .
>
> Best,
>
> [nickname]

The professor picked up on the clues, and accepted them, while still leaving both of us an out if there had been a misunderstanding ("if I may"). By using the more relational "Best" as a closing and signing only with a nickname, the professor was indicating a willingness to interact as colleagues rather than maintaining a hierarchical relationship.

Happy that I had accomplished both of my purposes, setting up a meeting and moving to a collegial relationship, my brief response was as follows.

> Dear [nickname],
>
> Thank you for the quick reply. I look forward to meeting you.
>
> Cordially,
>
> Gerry

By using only the professor's nickname in the salutation, and my own in the signature, with a relational closing, I acknowledged my acceptance of the first-name relationship.

The importance of cultivating relationships in academia cannot be overemphasized, even for scientists and engineers who often prefer to work with things and ideas. Collaboration is an inevitable part of modern research, and email correspondence plays a part in that.

17.4.2 Justifying your request

When it comes to writing the body of the email, use the same methods we will study for other academic writing. Begin by laying out the argument structure, then rewrite to make the argument more coherent and concise, rewrite again focusing on using words carefully to reduce the feeling of imposition, and finally proofread to catch grammar, spelling, or formatting errors. It would be ideal if you could model your email after successful examples. Unfortunately, such exemplars are harder to find than exemplar journal articles, but over time you can learn from emails sent to you. Writing textbooks can help if you want to improve more quickly.[5]

> A formal request is structured like an argument, so it is important to closely connect the request and justification.

It may seem strange to think of an email as an argument, but it is, with a request and justification instead of claim and support. Taking a letter of request as an example, the argument typically includes the request itself, a justification for the recipient's involvement, and an acknowledgment of the imposition, in some order. The justification can be made on different grounds, either personal or institutional. Moreover, because it is an imposition there needs to be room for negotiation and a reasonable timeframe to complete the request, with the final choice left to the addressee.

As you begin writing an email request, think first about your argument. What is the strongest justification you can make for their involvement? The personal request, "Since you are my advisor, can I talk with you about what courses I should take next semester?" sounds like you are placing them under obligation based on their position, and may not be well received. It would be far stronger to base the request on institutional policies, for example, "Since the university requires that I consult with my advisor before signing up for classes, would it be possible to meet sometime next week to go over the options?"

The justification should be appropriate to both the request and the relationship between sender and recipient. If you do not have an ongoing relationship, as with an advisor or coworker, is there any other relationship you could mention that would strengthen your case?

For example, when writing the first letter to the professor mentioned above to ask for a meeting, since I had no *personal* connection, following the request in the first paragraph I highlighted my connection with the *school* the professor teaches at and *acquaintance* with her colleague.

> My wife and I spent a sabbatical year at [professor's university], where I [goal] while she worked with [professor's colleague].

I added that I was using a *research* method promoted by the professor.

> Following a discourse analysis approach, as described in your various books . . . , I discovered that engineering articles typically do not follow the standard IMRD format, but an analogous format that I call IPTC.

Finally, I stated that I would be presenting at the same conference where the professor would be giving the keynote speech, to cement my relationship with the *field*.

> I will be presenting the format and the way I teach it during one of the concurrent sessions at [conference], as well as participating in an invited panel . . .

To minimize imposition, I left the timeframe open, both as to length and timing.

> Please let me know if you would be able to spare a few minutes sometime during or around the [conference] to talk with me about these questions.

Because I hoped to meet and talk with this professor, I gave more detail about myself than I would have if I were merely requesting information. The following example, written from a professor in Taiwan to a distinguished emeritus professor in forensic linguistics, shows how the same elements can be combined in a single paragraph. The first sentence shows the two are in the same *field*, the second talks about having taught at the same *school*, while the third points to common *research* interests. These, along with common *acquaintances*, are some of the most frequent means of establishing relationship. The request is found in the last sentence.

> I'm a sociolinguistic variationist, who organized the [conference] in Taiwan. I also taught at [college where distinguished professor once taught] from [dates]. As I have taught courses related to forensic linguistics during the past 6 years at [university in Taiwan], I'm working on a government book project proposal with the goal of purchasing 2,000 books in forensic linguistics to enrich Taiwan's research and teaching abilities in this field. Would you be willing to serve as one of our consultants to ensure the titles of the books we are planning to purchase are really the most important ones?

The justification can be stated before or after the request, depending on the recipient and the length of the email.

This brings up another question about argument structure. Should the justification be stated before or after the request?[6] Like all arguments, although there is no fixed pattern, there are common features and, as with journal articles, the expected order is different in Eastern and Western writing styles. In American style, it is common to make the request first, then build the argument to justify that request, as I did in the email just cited. In Asian style, the opposite is the case, with the justification presented before the request, as in the second example above.

Since the second example is a brief email, only one paragraph, placing the request at the end is not a problem, but if a longer justification is required it is better to place the request earlier. Busy academics may not read a long email unless the purpose is clear within the first paragraph. By analogy, in a longer email the first paragraph serves the purpose of an Introduction. Just as the research goal is found at the end of the Introduction of a research article, the purpose of an email is often placed at the end of the first paragraph of an email.

When making a request, try to put yourself in the position of the recipient. How would you feel if you received a similar letter? How much time will it take for the recipient to fulfill your request? How much lead time are you giving them? Make sure that the timeframe is appropriate to the work involved. Emphasizing your need, as in "Since the deadline is tomorrow, could you please look at this and return it to me with comments immediately," only works if you are in a position to impose on a subordinate, and is still resented. Apologies, such as "I am so sorry to get this to you so late, but if you could find the time I would really appreciate your comments," are more likely to elicit a favorable response. Remember that your recipient already has other things planned and is trying to fit one more thing into a busy schedule. Why should they do that for you? It is important to be as diplomatic as possible, and remember that it is a request, not a demand—the final decision is in their hands.

17.4.3 Amount of information

Many academics now receive tens if not hundreds of emails every day, so the amount of time spent on each must be limited. Just like an article, too little information will make it unclear, but too much wastes the reader's time, and makes it more likely that they will skim and perhaps miss pertinent information.

How much information does the recipient need? Enough to be able to respond appropriately. After you have decided what information to include, in what order, and written a first draft, it is time to rewrite for coherence and conciseness. Do not be like Blaise Pascal, who wrote, "I have made this letter longer than usual, only because I have not had the time to make it shorter" (Pascal, 1657). While that may work for letters written to a friend, it is not a good practice in academic email.

Consider first how much information to include about the relationship. If the recipient knows you well, it is not necessary to mention the relationship at all. If you are writing to a professor in a class you will be taking, it may be helpful to mention the connection, e.g. "I am signed up for your Engineering writing class." For most professors, your signature is all that is necessary after the first few weeks of class, unless it is a very large class. If you have met the person before, but have not seen them for a while, reminder of the context may be sufficient, e.g. "I don't know if you remember me, but we met when you spoke at [conference or location] last year." As discussed in the previous section, for an initial contact it may be necessary to establish a basis for relationship.

> You need to provide enough information for the recipient to be able to respond to the request, but not too much.

Next consider how much information they need to successfully comply with your request. Provide a direct statement of the purpose of the email, sufficient background information, and enough detail that they can act on your request immediately if they choose. If appropriate, additional information can be attached as a downloadable document rather than putting too much detail in the email itself.

One of the most common email requests graduate students write is to be excused from class. Which of the reasons shown below do you think would be most appropriate when asking to be excused from class?

"I cannot come to class next week . . ."

1 for personal reasons.
2 because I need to take my mother to the doctor.
3 because my mother has chronic arthritis, and can't get around by herself, so I have to go home to take her to see the doctor, and I won't be able to get a bus back in time for class.

Many students think choice 1 is sufficient, but most teachers prefer choice 2. As a teacher, I care for my students, and would like to know whether "personal reasons" means there is a problem, and if so whether it is likely to be short or long term, or whether it is something positive like a conference or job interview. Details may not be important unless I am your advisor, but knowing the general nature of the reason will determine whether an offer of assistance in making up the work or a message of encouragement or congratulation is appropriate. Similarly, in other request situations the goal is to provide the recipient with enough information that they can judge the importance and respond appropriately.

It should go without saying that the reasons given should be truthful and credible. If the recipient suspects either not to be the case, it is doubtful that the request will be granted. Thus, if the true reason sounds incredible, it may be useful to provide or offer to provide evidence to substantiate it.

17.4.4 Reducing imposition

Emails that impose on another person always run the risk of rejection. To minimize that risk, it is important to ask as politely as possible. This involves a second stage of rewriting, paying attention to connotation.

Compared to Asian languages, English is often viewed as being very direct. While it is true that in a face-threatening situation an English speaker will often speak directly to the other individual, instead of transmitting a message through a third person, as is common in Asia, the manner of speaking in that situation is often very indirect. This is illustrated in Figure 17.3.

Politeness in English is often achieved by indirect request. As an example, which of the following requests do you think is the most polite?

1 Could you please look over my paper and give me comments by next week?
2 I want you to look over my paper and give me comments by next week.
3 Please look over my paper and give me comments by next week. Thank you.
4 I would appreciate it if you could look over my paper and give me comments on it by next week.

Most students quickly eliminate choice 2, but some using English as a lingua franca believe that choice 3 is the best, because it uses both "please" and "thank you." Unfortunately, this form is viewed by native speakers of English as a veiled command rather than a polite request. Choice 1, an indirect request, is considered more polite, but choice 4, the most indirect, is considered the most polite, even though it lacks a "please."

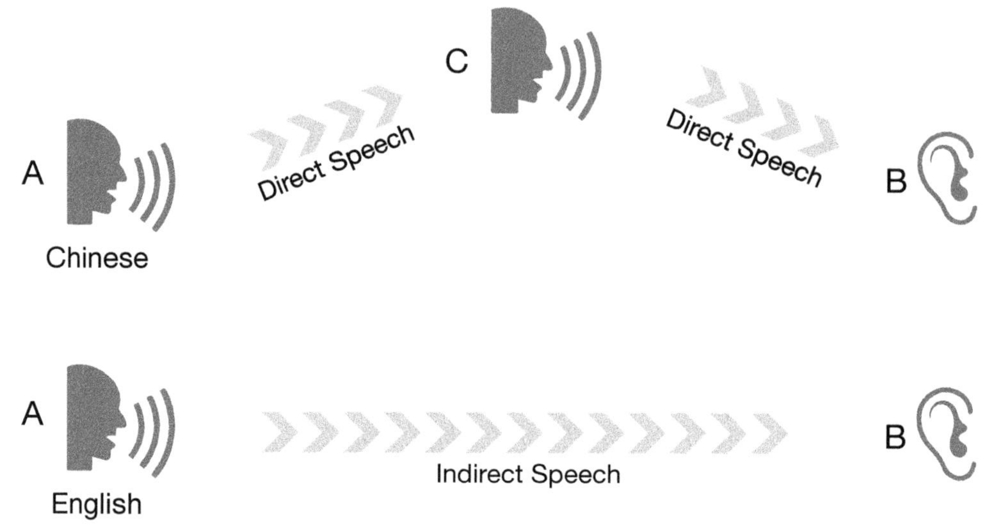

Figure 17.3 Indirect speech in Chinese and English.

In English, less direct requests are considered the most polite, leading to extensive use of modal verbs.

Some other ways of requesting are shown in Table 17.6, from most to least polite. Notice that all of the most polite forms utilize a modal verb (could or would), which has the effect of making the action potential rather than actual. The less polite forms use the modal "can," which can indicate ability as well as potential, or the word "hope," which conveys lack of certainty that you will help, although it still focuses on my need rather than your choice, as do those that are considered impolite.[7]

The choice of verb is also important. A Chinese student studying in the US pointed out that

> use of Want Statements along with the word "help" is likely to project a negative image as a needy, helpless student, which is quite the opposite from the attitudes such as independence and confidence that a graduate student is expected to demonstrate in U.S. academic culture.
>
> (Chen, 2006: 44)

Rephrasing your request to indicate your desire for guidance or advice conveys rather that you respect their counsel but are capable of independent action and decision.

17.4.5 Attention to detail

The final stage of rewriting, the one most speakers of English as a second language worry the most about, is probably the least important. In academia, more and more speakers of

Table 17.6 Request formats, from most to least polite

Most polite	I wonder if it would be too much bother for you to . . . I wonder if you could . . . I would appreciate it if you could . . . If possible, I would like you to . . . Would you mind . . .
Polite	Could you please be so kind as to . . . Could you please . . . Could you . . .
Less polite	Can you please . . . Can you . . . I hope you can . . .
Impolite	Please . . . I want you to . . . I need you to . . .

English as a first language are becoming aware of world Englishes and their counterparts in other countries who speak English as a lingua franca. As long as the email is comprehensible, correct grammar and spelling are far less important than the factors already mentioned. As with writing an article, correction of these small points of English usage should be postponed until the very end, after all the other factors have been considered.

> **Grammar and spelling are the least important aspects of email success in English as a lingua franca communication because they do not affect relationship.**

The reason that attention to English usage is the least important factor comes back to a theme repeated throughout this chapter: relationship. Salutation and closing are useful in negotiating relationship. Justifying why a person should be involved is important in developing or maintaining a relationship. Concern for the recipient, as shown in care for clarity, conciseness, and politeness, also contributes to positive relationships. As long as there is mutual intelligibility, correctness of form does not adversely impact relationship (think of all the errors in texting among friends). Since your time is as precious as theirs, use what time you have on things that promote not only understanding, but relationship.

Nevertheless, in some cases spelling may be important. One letter sometimes makes a huge difference. For example, I once had a student named Chester who mistakenly signed his name as "Cheater." Fortunately, the email was not about a question of academic honesty.

Evaluate your understanding

(1) What are the four basic principles of email etiquette?
(2) What factors interact in determining the level of formality appropriate for an email?
(3) What is meant by terms of address, and why are these important in email success?
(4) How can terms of address be used to negotiate a closer relationship?

(5) What might affect your choice of where in an email to state a request?
(6) How can you decide how much information to include?
(7) What is the simplest method for reducing the imposition of your request?
(8) Why are details like grammar and spelling the least important factor in email success when using English as a lingua franca?

Homework

(1) The next time you write a more formal email, apply the principles outlined above to construct your argument and rewrite before sending.

Hint: If you apply these principles in writing to a professor who has taught them in a writing class, it is even more likely you will succeed in your request, because they will know you were paying attention.

Teaching tip: Although this is not directly related to writing an academic paper, students often say it is one of the most useful lessons in the course.

Notes

1 One very common question from those using English as a lingua franca is whether email is countable, i.e. whether to say "I got a lot of email today" (collective) or "I got a lot of emails today" (plural). Either is acceptable, depending on whether you want to talk about volume or number.
2 My interest in email began while listening to a conference paper on email given by Yuanshan Chen at AILA 2011 in Beijing. Among other things, she said that one cause of email failure was "insufficient lexical and syntactical mitigation devices," a great phrase for linguists, but incomprehensible for others. While the following discussion is based on many pragmalinguistic and sociopragmatic principles, it is written for scientists and engineers, thus the focus is on the mechanics of writing rather than linguistic or communicative principles. For an excellent literature review and longitudinal study of email use by a graduate student using English as a lingua franca, see Chen (2006).
3 These have been distilled from various literature on interpersonal communication and email, as well as personal experience. While related to general communication principles identified by various authors, they are not specifically based on one model. An earlier outline of this section was presented as an appendix in Rau and Rau (2016).
4 Many other variants could be included and are easily found with a quick Internet search, including specialized closings such as "hope this helps," "looking forward to seeing you," or "in haste."
5 For example, Swales and Feak (2000: 228–256).
6 Request first is called deductive; justification first is called inductive.
7 Linguists call the first "Query preparatory" and the second "Want statements."

References

CHEN, C.-F. E. 2006. The development of e-mail literacy: From writing to peers to writing to authority figures. *Language Learning & Technology*, 10, 35–55.
PASCAL, B. 1657. *Lettres Provinciales*, No. 16.
RAU, D.-H. V. & RAU, G. 2016. Negotiating personal relationship through email terms of address. *In*: CHEN, Y.-S., RAU, D.-H. V. & RAU, G. (eds.) *Email discourse among Chinese using English as a lingua franca*, Singapore, Springer.
SWALES, J. M. & FEAK, C. B. 2000. *English in today's research world*, Ann Arbor, MI, University of Michigan Press.

Part 4

Creating your masterpiece

Chapter 18

Writing in stages

18.1 The hardest lesson for beginning writers

This chapter marks a major turning point in the book. In Part 1 you learned that argument is fundamental to academic writing. In Part 2 you examined research articles in your field to determine the argument structure. In Part 3 you discovered how arguments differ depending on the genre. Now you will apply all that information to your own writing. Since most of you are students doing some sort of academic writing, that will be the focus, but the same principles apply to technical writing. The method I am suggesting involves prewriting, writing, and extensive rewriting. You may wonder if all that is really necessary. I can sympathize.

It is somewhat ironic that I have become a writing instructor, let alone author of a textbook on academic writing. In high school and college I wrote research papers by making notes on notecards, sorting the cards into order (that was my prewriting), and typing the final draft directly (the day the assignment was due), without a rough draft or any rewriting. We called it keyboarding. Unfortunately, that did not work when I got to graduate school, and it is not an option for you.

Will you actually have to do extensive rewriting? In a word, yes, and the amount of rewriting required will continue to increase. Papers written for class are read by one person, the professor, and not rewritten after submission. A thesis will be read by several professors, each of whom will require changes before it is accepted. A published article will be read by your advisor, coauthors, reviewers, and the editor, all of whom will require changes after you think it is perfect. Moreover, it may be read by hundreds or even thousands of people, who will form their impression of you and your research based on that one paper. You need to make sure it is good. The amount of rewriting required is perhaps the hardest lesson for beginning writers to accept.

> You will have to do extensive rewriting to meet the demands of coauthors, reviewers, and editors, as well as the needs of your audience.

18.2 Why prewrite?

You may wonder, if you have to rewrite so much anyway, why spend time prewriting? Although rewriting is inevitable, prewriting can minimize the scale. There is a big difference between adding another reference, modifying the description of a process, and having to totally rewrite an entire article because it was poorly organized.

Prewriting is important for another reason: it helps overcome the inevitable writer's block that comes when beginning a large project. According to English author Neil Gaiman, "Being a writer is a very peculiar sort of a job: it's always you versus a blank sheet of paper (or a blank screen) and quite often the blank piece of paper wins" (Gaiman, 2004). The purpose of prewriting is to reduce the chance that the blank screen wins.

> The goal of prewriting is not to figure everything out, but to get you started—more like an artist's sketch than an architectural blueprint.

The goal of prewriting is not to figure everything out, to know exactly what the final product will look like, but to get you started in the right direction. It is not so much an architectural blueprint as it is an artist's sketch. This is very different from the days before personal computers, when retyping was time-consuming and costly, so minimizing revision was essential to success.

Perhaps it would help to give another analogy: writing is like putting together a 5000-piece jigsaw puzzle. Although strategies differ, most people first separate the puzzle pieces according to their main colors or other common features, while at the same time identifying the edge pieces. Once the edge pieces are in place, they begin to build inward from some prominent feature, some distinctive color. Undoubtedly, some pieces of that color belong in a different part of the puzzle, so when you get stuck, switch to a different color or part of the puzzle. Follow the main lines in the puzzle first, then fill in the gaps around them. Join pieces that fit together even if you don't yet know where the section belongs. The more pieces you put together, the easier it becomes to see the whole picture and decide where the rest might go. The larger the puzzle, the more the initial sorting helps.

> The writing steps suggested in this book are a scaffolding to help you get started, not a required pattern to follow forever.

Times and methods change, so you should feel free to adapt the procedures I suggest to your situation and work habits. Nevertheless, taking time to prewrite will save you time in the long run, as will following each of the rewriting steps in order. As you become a more experienced writer, you will be able to reduce or combine some of the steps. They are a type of scaffolding to help you learn to write, not a fixed pattern to follow rigidly forever.

18.3 Write as you go

As a writer myself, I have found one of the most important principles of writing is to write as you go. Whenever an idea comes to you, capture it. Novice researchers tend to think that they must have all the pieces in hand before they start writing. Actually, it will save time if you draft portions as you go, and it may be easier not to write the paper in its final order.

The two portions of a research article that graduate students find the hardest to write are the Discussion and the Introduction. Those writing in IMRD often have difficulty with the Discussion because there is no single correct answer. You are trying to compose an argument to convince your readers to see the data the way you do. That is the basic challenge of research. On the other hand, both in my own writing and my work as an editor, it is always the Introduction that needs the most revision. Even if you know the literature, it is not easy to decide how much to include or how to connect the various threads to create a cohesive picture.

The first Introduction or literature review you write will be the hardest. Another analogy may help explain why it is relatively easy for an experienced researcher to write an Introduction, but it is often the hardest division of the paper for a novice. Before you continue reading, think about the following question: how is writing an Introduction like giving directions to a new house?

The Introduction often requires the most revision, because it is the only division that does not describe your own work, but becomes easier as you know the field.

Consider the growth of your field of research as the growth of a town. Someone was the first to settle in that area, a pioneer. Many pioneers wind up living and working by themselves for their whole life, but if they find something valuable, others will start moving into the area and claiming a share of the land. Eventually a small town may be established, in which everyone knows one another. Over time, as more and more people move in, straight, wide roads are built to replace or connect the small winding paths that were there at the beginning. Everyone knows the founders, but gradually different neighborhoods take on their own character, and people generally stay within their own neighborhood. As the town gets more crowded, newcomers must either build on the edges or the few vacant lots in the middle. Later still, some of the old homes may be torn down and replaced by new development.

If you want someone to find your new house, you need to give them directions. You will not talk about every road in town; only those they need to follow. You may mention famous landmarks along the way by name, but in general you will just tell them what roads to turn on. The same is true in writing an Introduction. You do not need to mention every paper in the field—only those that lead to your present work, with special mention of landmark papers along the way.

The pioneer has seen all the development, seen every road as it is built, knows where each leads and who lives on it, and has met each person as they move into town. As a newcomer you may have no idea where the various roads lead, or who lives on them. You may have a map with street names, but do not know where the landmarks are or what they look like. Nevertheless, the longer you live in the town, the better you know it. When a new building springs up, you see it being built and begin to talk about it before it is even finished. You know various routes to any location in town. The same is true with an academic field. The longer you are there, the easier it will be to direct people to your work.

Next, think about what you encounter when you read a research article, and when you would be able to write each. The first thing you see is the Abstract, but since that is a summary of the article, it cannot be finalized until the article is complete. The first part of the article itself is the Introduction, but as I just said, that is one of the hardest sections of the article to write. Moreover, the purpose of the Introduction is to point to a research gap and show how your work is going to fill it, which means you need to have done that work and be sure you can indeed fill that gap in previous research before you can write it.

> Writing up your work as you do it is easier than trying to remember or decipher your notes six months later.

But why wait until the work is complete to start writing? When are you going to be most clear about what methods you used—as you do them or six months later? Why not write it as you do it? Undoubtedly you will change the wording later, but at least you won't be struggling to remember what those cryptic notes mean that were once so clear.

In practice, this means writing the Methods or Process division first, as you perform the work, followed by the Results or Testing. Since the analysis depends on the research questions asked, it is beneficial to write the Introduction next, to make sure you are clear what the goal is and what has already been done, before returning to write the Discussion and Conclusion.

One other thing I find helpful when writing is to keep at least two files open: the section I am writing at the time and another file with an outline and notes on the whole project. As I am writing one chapter, ideas will come about how to reorganize or what to add to another chapter. When the idea comes, I simply switch to the other file, record the idea before I forget it, then go back to my main document.

> The best wording may occur to you while writing something else, walking across campus, or when you wake up. Be ready to capture it.

Furthermore, when away from the computer, I keep some way of recording notes handy. In the past, that was a pen and paper. Now the omnipresent cellphone fulfills the same purpose. Whenever the Muse speaks, I am ready. Frequently the best wording comes at an odd moment, walking across campus or lying in bed, but if you do not record it then, it will be gone forever. As best-selling author Saul Bellow said, "You never have to change anything you got up in the middle of the night to write" (in Bear, 1992: 93).

18.4 The 7Cs of Change

I propose a method of rewriting I call the 7Cs of Change.[1] A summary of the 7Cs is shown in Table 18.1. The first two Cs (Coherence and Conciseness) constitute rewriting at the

Table 18.1 The 7Cs of Change

Type of change	*7Cs*	*Purpose*	*Level*	*Scope*
Rewriting	1 Coherence	Organization	Paragraph	Section
	2 Conciseness		Sentence	
	3 Connection	Clarification	Sentence	
	4 Connotation		Words	
Revising	5 Consistency	Finalization	Words	Whole paper
	6 Correctness		Format	
	7 Collaboration		All	

level of the paragraph or sentence to ensure your argument structure is clearly organized and eliminate unnecessary material. The third and fourth (Connection and Connotation) involve restructuring sentences and choosing words to further clarify your argument. The last three (Consistency, Correctness, and Collaboration) comprise revision,[2] looking at the whole document again to check for consistency of word use, spelling, formatting, and other such issues, or seeing it from a new perspective after asking for comments from coauthors, making sure your argument is not misunderstood.

Inexperienced authors make relatively few changes themselves during rewriting. As a result, they need to make extensive changes at the level of collaboration, after their advisor or coauthor reads it. Experienced authors, on the other hand, learn to make large organizational changes before working on smaller ones like word choice. This reduces the total work since moving or adding even a single sentence requires rewriting several sentences around it. We will examine the 7Cs individually in Chapters 21–23.

It makes sense to get the organization right before working on details, because every change in order requires rewriting several surrounding sentences.

18.5 Do it right the first time

Some things are worth doing right the first time, as it will save considerable time in later revision. This includes formatting problems created by our dependence on computers and some common punctuation errors.

Some publishers provide templates for submitting articles. These templates have sample sections, paragraphs, tables, and figures. Once your text is ready, you simply delete their text, insert your own, and embedded codes ensure it is properly formatted. Many universities have standards for formatting theses and

Small formatting and punctuation errors are easy to avoid but take time to correct later.

dissertations, specifying how paragraphs are to be formatted or the number of spaces following a period, but lack such a template. By checking the requirements before you start to write, you will avoid many tedious small changes later.

Other problems with formatting arise from computer defaults. When you copy and paste in MS Word,[3] the default is "keep source formatting," which may differ from your document. The difference between Times New Roman and Cambria or between size 12 and 11 font may not be obvious (did you notice that each was written in the respective font/size?), but the editor will ask you to change it. Using "keep text only" when pasting plain text into your document will save time later.

> Some common problems arise from pasting from other documents, others from foreign language entry systems.

One formatting issue involves line spacing for Chinese, Japanese, and Korean. Lines copied from such documents, even if typed in English, will retain the wider line spacing designed to accommodate characters. To correct this in MS Word, uncheck the box on the paragraph settings to "snap to grid when document grid is defined."[4]

Another problem for Asian users is the difference between typing English and full English entry. In the former all the punctuation marks, parentheses, and other marks will appear in Asian format rather than English. The difference is illustrated in Figure 18.1.

The shading, none of which include spaces, illustrates the problem. In Chinese a punctuation mark takes the same amount of space as a character, but in English it is smaller than the average letter. Thus a Chinese parenthesis appears to have space both outside and inside which is part of the parenthesis character itself. Similarly, periods and commas look like they have spaces around them when they really do not. Chinese writers will often either follow the same entry in English(not leaving a space outside the parentheses)or try to make it look like what they are used to seeing in Chinese （adding spaces both inside and outside the parentheses）. Nevertheless, when in America (or Britain) you must do as the locals do, with no space inside the parentheses but a space before and after them (unless followed by punctuation). The examples in the previous sentence can serve as a model of proper English usage.

In an English font, punctuation is at the bottom of the line and takes a small amount of space (and there is space around the parentheses only if you specifically add it). The line spacing is noticeably narrower than Chinese single line spacing.

In a Chinese font that has defaulted to English，punctuation may be in the middle of the line and take as much space as a character（like this）so there appears to be space around the parentheses or after punctuation when there really is none。The line spacing is also wider, to leave vertical space for Chinese characters。

Figure 18.1 English vs. Chinese punctuation and line spacing.

Exercise 18.1 Instructions to authors

(1) Check the formatting guidelines for your writing. Ask your school for their thesis/dissertation guidelines or look on the journal homepage for instructions to authors, author guidelines, submission guidelines, or some similar link.

Hint: We will refer to the instructions to authors several times during the rewriting process. File them for future use.

(2) Look through these guidelines to determine their requirements for font, font size, paragraph formatting, spaces after periods, British or American English, and any other general formatting features that affect the whole paper.

Hint: Do not worry about headings, tables, and figures (which will be a large part of the instructions). We will come back to these in Chapter 24.

(3) If your school or journal does not provide a template, create your own template, incorporating the font, font size, and paragraph formatting specified.

Hint: This may seem like a waste of time, but it takes far less time to make a template and use it consistently than it does to reformat every document.

18.6 Avoiding plagiarism

One final general issue we need to consider is plagiarism. Novice writers often want to know what constitutes plagiarism and how it can be avoided, particularly when citing numerous works in the Introduction.

Plagiarism is a particularly Western idea, based on the ideas of individualism and individual ownership, and therefore hard for Eastern cultures to understand. In many Eastern cultures the teacher is the expert. To quote their answer verbatim is a sign of respect. Memorization is highly valued and often an answer must be quoted perfectly to receive credit. Knowledge, like many other things, is communal property. In Western culture, the reverse is true. The teacher is a facilitator, encouraging students to think critically and form their own opinion. Quoting verbatim without citing the author is plagiarism, and the value of direct quotes is minimized, as it hinders self-expression. Memorization is not valued in the school system, with test questions worded to ensure that even if students memorize the entire textbook, they can only get full credit if they can express the answer in their own words. Knowledge, like other commodities, can be copyrighted and individually owned (intellectual property rights).

> **Plagiarizing one time can ruin an academic career.**

Although the concept of plagiarism is culturally determined, the need to avoid it has become an accepted part of academic life worldwide. A student caught plagiarizing may fail a class. As an academic or professional it can lead to censure or dismissal. Thus when writing in English, Asian students (and those from similar cultures) need to set aside their own thoughts on what is appropriate (just as they must set aside their own thoughts on what is an appropriate writing style) and follow English standards.

Plagiarism is defined as using the words or ideas of another person as if they were your own.[5] In other words, you need to either (1) quote the person directly, (2) change it slightly and cite their work, or (3) change it enough that it is no longer recognizably theirs. Direct quotes are common in the humanities and social sciences, where whole paragraphs may be quoted, but it is rare to quote even a single sentence in the natural sciences and engineering. Since you want to relate your work to that of others, you cannot make it unrecognizable. That leaves the second option. You must learn how to change the wording and cite the previous work. But the question is, how much change is necessary?

To illustrate what is considered plagiarism, in Table 18.2 compare what Swales and Feak (2004) wrote, quoted directly in the second column, with what I have written in the last column. Which of the examples would you consider to be plagiarism? Note that each of my rewrites attempts to illustrate the principle Swales and Feak mention.

The amount you can copy from another work without citation varies by field and division of the paper. Rewriting several times helps avoid plagiarism.

If you think that the first two are plagiarism, but the latter two are not, you would be correct in most cases, although the amount of similarity allowed varies somewhat by field. In highly technical fields like computer science it is not uncommon to reuse whole

Table 18.2 Examples of plagiarism and non-plagiarism

Modification	*Swales and Feak (2004: 173)*	*Plagiarism or not?*
Substituting	"Copying a paragraph making only small changes, such as replacing a few verbs or adjectives with synonyms."	Copying a paragraph, but making a few minor changes, like changing some of the verbs or replacing adjectives with other words meaning the same thing.
Reordering	"Cutting and pasting a paragraph by using the sentences of the original but leaving one or two out, or by putting one or two sentences in a different order."	Putting one or two sentences in a different order, or leaving one or two out, but basically cutting and pasting a paragraph from the original.
Rewriting	"Composing a paragraph by taking short standard phrases from a number of sources and putting them together with some words of your own."	If all the authors in your field use the same standard phrases, you will need to use them as well, but by putting them in the context of your own work you make them your own.
Paraphrasing or expanding	"Paraphrasing a paragraph by rewriting with substantial changes in language and organization, amount of detail, and examples."	If you totally reorganize a paragraph, making it longer by adding examples as I have to the categories suggested by Swales and Feak (2004) or summarizing points to make it shorter, this is considered paraphrasing.

Adapted from Swales and Feak (2004).[6]

sentences from previous works without citing the original source.[7] Method sections or other highly condensed writing may be essentially equivalent apart from the different variables or materials used.

The American author Flannery O'Connor said, "There may never be anything new to say, but there is always a new way to say it" (O'Connor, 1969: 76). Nevertheless, in practice, how can you avoid plagiarism? The best method I have found is to read your source at least twice, to make sure you understand the content. After that, make notes in your own words, retaining only the technical words that cannot be changed. Use phrases rather than full sentences and try to write things in an order that makes sense to your own research, rather than the order of the original paper. Finally, close the source and write your summary or paraphrase, preferably at least a few hours after making the notes. Most importantly, write in stages, as I will describe in the following chapters. If you follow this procedure, your final version should have been rewritten enough times that plagiarism is not an issue.

Exercise 18.2 Paraphrasing practice

Read one portion of one of your exemplar articles. Set it aside and write one sentence summarizing that portion. Do this for each portion of a division. When you read these sentences in order, it should reveal the structure of the argument in that division.

> *Hint: Choose portions longer than a paragraph, perhaps related to one component or test, which often begin with sentences indicating a major transition.*

Evaluate your understanding

(1) What is the advantage of prewriting?
(2) What are two advantages of writing as you go?
(3) Why does it make sense to rewrite several times, beginning with the largest changes?
(4) Why should you pay attention to some small details such as punctuation as you write?
(5) What is the best way to avoid plagiarism, and why is it important?

Homework

(1) Set goals for finishing different stages of your writing, to make sure you will be able to finish by the deadline.

Hint: Allow enough time for rewriting. I suggest the following as a rough guideline, based on the percentage of time available before your deadline:

10% Prewriting (argument outline, brainstorming, linearization), Chapter 19
40% Writing, Chapter 20
40% Rewriting (Cs 1–4), Chapters 21–22
10% Finalization (Cs 5–7, graphics, references, citations, title, abstract), Chapters 23–26

> *Teaching tip: Setting deadlines for different stages of writing provides a scaffolding for students to follow when they do their own writing later and ensures that the work is not all done at the last minute.*

Notes

1 I am not the first to come up with a writing framework based on 7Cs. John Blake called my attention to at least three previous schemes: The Seven C's of Effective Communication (www.managementstudyguide.com/seven-cs-of-effective-communication.htm), The Seven Cs of Business Letter Writing (http://epi.sc.edu/~alexandra_rowe/FOV3-001000D5/S002A8538), and one presented in an old Hong Kong business writing book entitled *Reader-friendly writing* (out of print). LaPlante (2018) suggested 5Cs of Technical Writing. My 7Cs of Change were developed entirely autonomously, before hearing of any of these.
2 In American English rewriting and revision are often used synonymously, but I am separating the ideas of rewriting (literally, writing again) from revision (literally, looking again). In British usage, to revise means to look over material again, often in preparation for an exam, which is called review (another way of saying looking again) in American English.
3 Throughout the book I will use MS Word as an example of a word processor, because it is by far the most widely used. Terms differ in other programs. LaTeX is used extensively in computer science and mathematics but is rare in other fields.
4 This option is not visible in English versions of Word unless one of the Asian language packs is also installed. Because both are "single" line spacing, although the actual spacing differs, selecting single line spacing does not work.
5 If a publisher holds the copyright, you cannot even copy your own words without permission. Thus you cannot publish material that is substantially the same in two different journals, even in a different language.
6 The original list also includes direct copying and quoting, neither of which involve modification.
7 John Blake (personal communication).

References

BEAR, J. 1992. *The #1 New York Times bestseller: Intriguing facts about the 484 books that have been #1 New York Times bestsellers since the first list in 1942*, Berkeley, CA, Ten Speed Press.

GAIMAN, N. 2004. On writing. *Neil Gaiman's journal* [Online]. Available from: http://journal.neilgaiman.com/2004/02/on-writing.asp.

LAPLANTE, P. A. 2018. *Technical writing: A practical guide for engineers, scientists, and nontechnical professionals*, Boca Raton, FL, CRC Press.

O'CONNOR, F. 1969. The nature and aim of fiction. *Mystery and manners: Occasional prose*, London, Macmillan.

SWALES, J. M. & FEAK, C. B. 2004. *Academic writing for graduate students: Essential tasks and skills (Vol. 1)*, Ann Arbor, MI, University of Michigan Press.

Chapter 19

Prewriting

19.1 Planning your masterpiece

In the first chapter I compared learning to write to learning to paint. An experienced artist may paint on a blank canvas, but for a beginner it is useful to sketch out your masterpiece before picking up the brush. In this chapter you will learn how to sketch the broad outline before beginning to write, modeling successful exemplars, if available. This is not a paint-by-number. You are not trying to get every detail in place first, then just fill it in. Rather it is a quick sketch to make sure you know what the focus of attention is, what other elements you plan to include, whether each will be foregrounded or backgrounded, and how they relate to one another.

> Prewriting does not specify every detail, but helps you decide what the focus of attention is, other elements to include, and how they relate to one another.

This chapter is not about prewriting in general. There are many good books and websites on academic writing, so I will not duplicate their efforts.[1] Rather, I will tie the basic techniques in with the principle of component analysis. My goal is not to explore all possible methods of prewriting, but to encourage you to see its importance and find a method that works for you.

19.2 Initial sketch: argument structure outline

In Chapters 11–15 we discussed many genres of writing, how they differ in goal and audience, what components they are likely to contain, and how to find exemplars. Now it is time to lay out the overall structure of whatever you want to write.

> Lay out your writing based on the argument structure, rather than a topic outline by content.

I suggest you begin with the expected argument structure rather than a content outline. At this point you understand the argument structure of research articles, the general claims of each division, the component claims and where they are normally found, and the evidence used to support each claim. If you are writing a research article, you can use this information to plan your own writing. If you are writing something else, by comparing the typical argument structure of that genre with research articles you will be able to lay out other types of writing as well.

Before writing anything, the following steps should be taken:

1 Obtain examples of successful documents of that genre, if possible;
2 Determine the overall organization and structure;
3 Identify the main claims and components;
4 Identify what is used for support;
5 Speculate on reasons for the sequencing;
6 Evaluate which document is the best pattern to follow for your own writing;
7 Outline the argument structure for your own writing based on that pattern.

In Chapters 1–9 you learned how to do steps 1–6 for research articles, so let's proceed to step 7. If you are writing something else, adapt the outline to the structure appropriate to your genre.

Let's say that we want to write an article for the same journal and on a similar topic as our sample engineering article, "Spatial query integrity" (Hu et al., 2013). Based on what we found in previous chapters, we might lay out the structure as shown in Table 19.1. The section titles in quotes stand for titles specific to the new research, as opposed to the others which are general titles common in the field.

Remember that the division indicates the general claim supported in those sections, while the component tells us the specific types of claims and evidence needed in each. The length and number of citations will undoubtedly change, but give us a range to aim for.

Table 19.1 Argument structure outline based on the article "Spatial query integrity"

Div.	*Section title*	*Length (words)*	*Components and support*		*Citations*
I	1 Introduction	800–1200	1a 1b 2b 3b	I a I b I d V a	5–10
	2 Related Work	500–800	2b	I d	20–25
P	3 Preliminaries	400–600	4a	I b, c	0
	4 "My New Algorithm"	3500–4500	5d	II e	0
	5 "Using the Algorithm"	800–1200	5c	II d	0
T	6 Experiments	1500–2500			
	6.1 Experimental Settings		6a	II f, I d	0–5
	6.2 "New vs. Old Algorithm"		7a, 8e	II g, h, i	5–10
	6.3 "Practical Application"		7a, 8d	II g, h	0
C	7 Conclusions	100–300	(10a) 10b		0

Now it's your turn. First do steps 1–6 for whatever you want to write, so you have a good idea of the structure you are trying to copy. Use the information in Chapters 11–15 to guide your analysis, using exemplars from your class, school, or granting agency if available.

Next, begin your outline by writing tentative section titles. For example, for a research article in IMRD format, use the section titles that are standard in your field, whether IMRD, combined R&D with C, or whatever variation is common in your journal, such as "Experimental" in place of Methods. In IPTC format, begin with any section names that are the same or similar for most of your exemplar articles, then give rough names for those specific to your research. Don't worry about getting them exact—just an idea for now of how many sections you will need.

Next, think about how long each section should be. If this varies from one exemplar to another, give a range. Adjust the number of words devoted to each section, based on the focus of your article. If there is great variation in total words, is there a typical ratio you could use as a guideline for section length?

Now list the components and type of support expected for your genre. Parentheses can be used to indicate a component present in only some exemplars. Finally, if your document will need citations, give yourself a goal for each section, again based on the exemplars.

> This sketch can be used as you write to check if you have included everything you need, not too little or too much, and where you need to cite past work.

This completes your sketch. Remember that it is only a sketch and will undoubtedly be modified as you add details, but it gives you a starting point, general direction, and reminder of what is most important.

Exercise 19.1 Argument structure outline

For each step, follow the pattern of your exemplars, if available.

(1) Propose tentative section titles for your writing.

Hint: Should your titles be general, or specific to the research topic?

(2) Give a rough goal for the length of each section.

Hint: Adjust the relative length of the sections based on the focus of your writing, and remember that these are just rough estimates, not exact numbers you have to match.

(3) List the components and support you need to include in each section.

Hint: If all your exemplars include a component, you should as well. If only one or two include that component, put it in parentheses. Listing it will remind you to consider later whether it is important to your research.

(4) List an approximate goal for the number of citations you should have in each section.

Hint: This will remind you how many total articles you will need to read.

Teaching tip: This outline will also be useful to you as a way of keeping track of the goals for each student's writing if you are not personally familiar with the format and expectations of each genre or journal.

19.3 Adding the base layer: brainstorming

The next step of prewriting is brainstorming, getting as many ideas as possible down on paper so you have something to organize. Some of the ideas may turn out to be irrelevant to your topic, and later be discarded, but at this point you are trying to set forth as many ideas as you can that might be included and begin to think about relationships between them. Nevertheless, as Canadian writer and humorist Stephen Leacock said, "Writing is no trouble: you just jot down ideas as they occur to you. The jotting is simplicity itself—it is the occurring which is difficult" (in Leacock & Lynch, 2002: 194).

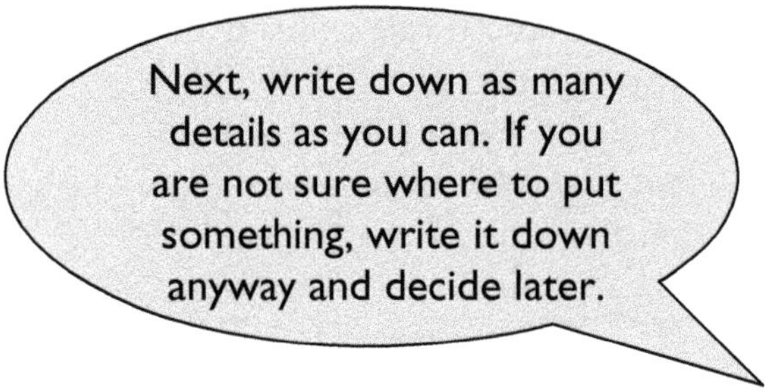

Thus I suggest that instead of brainstorming randomly, think about each component. For example, for Component 1, what have previous authors said about the importance of similar work? Write as many things as you can think of, then move to the next component. If as you are working on Component 2 you think about something else related to Component 1, go back and add it. Don't worry at this point about getting the ideas in order or how they relate to one another. If you are not sure which component something relates to, or if it might relate to more than one, just write it somewhere. You will organize in the next step. You can either set a specified time for brainstorming or just write until the ideas stop flowing.

As a visual learner, I like listing. I find it easier to do this on paper rather than computer, as it is easier to add things to different lists as thoughts come, without worrying about moving around the page on the computer, misspellings, improper auto-corrects, saving, or other computer-related issues. On the other hand, doing it on a computer makes reorganization afterward easier.

Auditory learners may prefer another method, sometimes called "freewriting," writing down all ideas as they come without trying to organize them. With modern technology, it is easy to do this by speaking into a smartphone and using a talk-to-text app to convert it into written text.

Kinesthetic learners may prefer to sketch out connections between various ideas, using words with circles around them for each main concept. Different variants of this have been called clustering, webbing, and mapping. This can also be used after listing or freewriting to show how ideas are connected.

At the same time, list papers you need to cite for each component. List the author, year, and key ideas from each paper that you want to include. If a paper relates to more than one component, list it in both places.

Next, identify which components you can already begin writing, which you could read more past research on to get the necessary information, and which will need to wait until you have done more research or analysis. This will help you plan which sections to write first and how to gather the other information you need.

> Now decide where to begin writing, based on what information you already have available, what you could get quickly, and what you will not have until later.

Currently we are considering organization of the whole paper, but you can do similar brainstorming later as you begin writing each section. You may also realize later that some of the items you listed should be eliminated, because they do not deal directly with the goal of this writing.

Exercise 19.2 Brainstorming

(1) For each section, consider the components you need to include, and write down as many ideas as you can that you think you will need to include in that section, using whatever method of brainstorming works for you.

Hint: Include papers you need to cite and relevant information from each.

(2) Identify which sections or components you can begin writing, which must wait until you are further along in your work, and what articles you need to read.

Hint: How many more articles do you need to read? Make a reading plan to finish them several weeks before the deadline.

19.4 Placing the main elements: linearizing

Recall that an argument can be presented in many different orders, which can be categorized as physical, temporal, and relational (section 9.5).[2] Look back at your exemplars to remind yourself what they use.

Before you decide what order to follow, it is useful to consider various possible orders, and different relationships between the past and present research that you could pursue. Not only does this have the obvious advantage of forcing you to think about the argument structure before you start writing, it also gives you other options you can return to if the first proves difficult. If you have not considered alternatives first, it may be more difficult to think of them after you start writing from one perspective, just as it may be hard to end a relationship, even if you know it is bad, unless you have a better alternative at hand.

> Recall the sequencing strategies used by your exemplars, but also consider other strategies that might be better suited to your topic.

In Exercise 19.3 you will make a more detailed outline using topic sentences to linearize your writing. Remember that this is still like a rough sketch or sorting puzzle pieces into piles. You will undoubtedly change the order and even the number of sections or subsections as you write, so do not worry about getting it perfect—you won't. As your writing ability improves, you will find that you have begun to internalize the expectations and can simply write.

Exercise 19.3 Linearizing your article

(1) After you have finished brainstorming, group related items together.

Hint: Think about different groupings that would work.

(2) Thinking about the argument structure you are aiming for (Exercise 19.1), order the groups of ideas within sections.

Hint: You may need to change your groupings or sections to do this. Remember that the order is still tentative and will probably be changed during the writing and revision process.

(3) Generate a content outline, with each primary heading representing the main idea of one section, and subheadings for paragraphs within that.

Hint: Copy the argument structure outline as a new file, eliminate the columns for length, components, and citations, and add lines for each paragraph.

Hint: Do not number the headings; just indent to show sub-categories. This will make it easier to reorganize.

(4) Think about how many paragraphs you will need to convey all the main points identified in your groupings. Write a topic sentence for each paragraph.

Hint: Use the component markers you found in your exemplar articles. For example, for Component 1, it is unlikely that any of your articles use the pattern, "My topic is important because . . . ." This may be the underlying claim, but not how it is stated.

Hint: Remember that reading the topic sentences should allow you to follow the basic argument, but not all the details and support. Do not worry about getting the wording perfect but be sure the order makes sense. If not, reorganize or rewrite.

(5) Now think again about your argument. If you see gaps, add topic sentences to fill them, or placeholders to indicate what you will need to add later.

Hint: If you do not yet have the information necessary to fill a gap but know there is something missing (e.g., citations or a test you have not done yet), use a placeholder, for example [data, test 1].

Evaluate your understanding

(1) What is the purpose of prewriting?
(2) How does creating an outline based on argument structure of exemplars differ from a traditional content outline?
(3) What methods can be used for brainstorming?
(4) Why is it useful to consider various organizational strategies before beginning to write?

Homework

(1) Ask a peer to look over your list of section titles and topic sentences from Exercise 19.3, to see if the logic of the argument is clear, and revise if necessary.

Hint: When you write, you can start with these topic sentences if you are writing in extended style or replace them with the detailed contents of the paragraph if writing in condensed style.

Teaching tip: The purpose of writing topic sentences instead of a phrase outline is to help clarify how the ideas tie together. Details can be listed as points below each.

(2) For academic writing, write a preliminary research goal, title, and abstract.

Hint: These will undoubtedly be very different when your article is complete, but like prewriting they help ensure you know the overall goal and direction.

Supplemental material: Class exercise 19.1—Linearizing practice

Notes

1 One of the best I have seen was developed by The Writing Center of the University of North Carolina at Chapel Hill. Several very useful handouts are available for download, including brainstorming, argument, evidence, and thesis statements, at http://writingcenter.unc.edu/handouts.
2 These correspond to three perspectives listed in the brainstorming handout from UNC-CH: describe it, trace it, and map it, respectively. The handout also discusses other techniques that seem to be more useful for social sciences and humanities than for science and engineering, but may be helpful for some writers.

References

HU, L., KU, W.-S., BAKIRAS, S. & SHAHABI, C. 2013. Spatial query integrity with Voronoi neighbors. *IEEE Transactions on Knowledge and Data Engineering*, 25, 863–876.

LEACOCK, S. & LYNCH, G., ED. 2002. *Leacock on life*, Toronto, University of Toronto Press.

Chapter 20

Writing

20.1 Just do it

When the time comes to write your paper, the best advice is, to borrow an overworked phrase, "Just do it." All writers experience writers' block at some point, not only when it is time to start a new project, but throughout the process. The difference between those who become good writers and those who don't is that the former do not quit. Just like any other task, the only way to succeed at writing is to sit down and do it, whether you feel like it or not—every single day. Best-selling British author Philip Pullman is widely quoted as saying,

> Do plumbers get plumber's block? What would you think of a plumber who used that as an excuse not to do any work that day? The fact is that writing is hard work, and sometimes you don't want to do it, and you can't think of what to write next, and you're fed up with the whole damn business. Do you think plumbers don't feel like that about their work from time to time?
>
> (Pullman, 2017)

> **Do not wait until you feel ready to write; just begin. If it is bad, set it aside and start over. Working on different parts at the same time helps you work effectively.**

In the first part of this book you learned to read articles in your field to see how others do it. The more you read, the more ideas you will pick up, not only about the subject but also about how to structure your writing. But the time comes when you just need to sit down and write. After you write, look at it, and if it is bad don't be afraid to throw it out and start over. After two or three tries you will find that it looks a lot better than it did the first time.[1]

Another tip is to write several things at one time, whether different parts of one project or different projects. All writers reach a point where they run out of ideas. Sometimes it is necessary to just plow through, no matter how you feel, but effective writers will switch to another section or a different task, rather than either sitting there frustrated or quitting. Perhaps you will find that you are more effective at writing in the morning and rewriting in the afternoon, or conversely that you need to do the final revision when you are fresh but can write a rough draft later in the day. Maybe you can work on details like formatting when you are tired of both writing and rewriting. You need to learn what works best for you, but no matter what, keep going.

20.2 Keeping track of where you are

The next three sections offer practical hints to help you write more effectively, beginning with moving between different sections of a document quickly without scrolling so you don't lose a thought before you find where you want to write it. In MS Word, the navigation pane helps you do this.

To use the navigation pane, format your document using "Styles." Begin with your prewriting from Exercise 19.4. Make sure you have topic sentences or ideas under each title as a placeholder for the plain text you will type for each section. Then highlight each section title and click "Heading 1." Do the same for any subsection titles, clicking "Heading 2."

> Format your document using headings, especially for longer documents, so you can move quickly from one section to another without scrolling.

Now, on the "View" tab click the box to show the "Navigation Pane." It should appear to the left of your document, showing your titles. To move from one section of the document to another, simply click a title. If you add section titles later, format them the same way and they will immediately be added to the navigation pane.[2]

20.3 Keeping track of citations

One issue that is somewhat unique to engineering is the question of how to manage citations while you are writing. In most engineering articles, citations are indicated by number rather than the author name and date format common in science. Since those numbers are based either on the order of appearance in the text or an alphabetical list of authors, they will change as you add more references.

If you already know how to use a citation manager, feel free to use it as you write. If not, I recommend that you first insert the author name and date in parentheses as a placeholder, then insert the actual citation marks in the final stage (section 25.3). This has at least four advantages:

> Use placeholders (author-date) for citations while writing, to become familiar with the references you are citing and avoid confusion.

1 You become more familiar with the authors and dates of publication of important works in your field. You will need to refer to them that way in presentations;
2 The reference numbers will not change while writing, eliminating confusion;

3 You will not be distracted by entering information in the citation manager;
4 Citation managers add a level of complexity to word-processing programs, making errors more likely, so delaying entry eliminates a source of frustration.

20.4 Keeping track of your work

One final question is how much of your writing to save. In days gone by, writers kept multiple versions of their work, from handwritten outlines and drafts to various typed and marked-up copies, in addition to the final draft. Historians now plumb those records to trace the development of thought of the author. Nowadays, although storage of multiple versions is much easier, it is also far easier to simply hit "save" and overwrite the previous version. Nonetheless, there are benefits of saving multiple versions.

Hopefully you already make a practice of regularly backing up your files and keeping the backup in a separate location. This can avoid the tragedy of being just a few months away from getting your doctoral degree and losing all your research and writing in a fire.[3] The rationale for saving multiple versions is the same. If one file becomes damaged or lost, you do not have to start again from scratch. This needs to become part of your regular routine.

A second advantage is that having multiple copies encourages you to try different ways of writing the same thing. If after looking at your writing you think it is trash, don't trash it completely, but save and close it, open a new file with a new filename, and start over. If the new version is still no good, do the same thing, reorganizing your outline or notes to try a different approach. Eventually, within three or four tries, you will find something that flows fairly easily. Even at this point, don't trash the other versions. As you continue to write you may remember that there was one line you wrote before that really sounded good and you will be able to go back to find it.

> Keep multiple versions of your writing, in multiple locations, both for safety and in case you change your mind, but make sure you only work on the most recent.

Multiple versions can be designated by placing a version number or date after the main filename. Move inactive files to an archive folder, to make sure you do not open and work on the wrong version by mistake. Whether you delete that archive when you are done is a personal choice.

Finally, learn to use "Track Changes" in MS Word. Many editors now comment on and change documents electronically using this feature. As the cloud becomes more accessible, it is likely that platforms such as Google Docs that allow multiple users to comment on or change the same document in real time will become more popular. Since technology is changing so quickly, I will not dwell on the mechanics of these features, but you will need to learn whatever is commonly used in your department and field.

This chapter has been brief, because at this point you do not need to read more, but to begin writing. You may use any of the methods described above to assist you, but at some point the only solution is "Just do it." But remember, the rough draft will not be perfect. It may not even be that good. As Ernest Hemingway charmingly put it, "The first draft of anything is shit" (in Samuelson, 1984: 11).[4] In the next few chapters we will learn how to transform that into useful and inoffensive fertilizer that will foster the growth of new ideas.

Evaluate your understanding

(1) How does working on different parts of a large project at the same time allow you to work more effectively?
(2) What is the advantage of using the navigation pane?
(3) What are the benefits of using placeholders for citations as you write?
(4) Why is it important to keep multiple versions and backups of your work?

Homework

(1) Write a rough draft of the first section of your writing.
(2) When you are done, ask someone to comment on it—your advisor, writing teacher, or a fellow student.

Hint: Be prepared for criticism. Be prepared for marks all over the page and corrections on every line (it may not be that bad, but be prepared!). Remember that you are not likely to do any better with your first draft than Hemingway did. You will learn how to improve it step by step in the following chapters.

Teaching tip: Do not assume that students know how to use features of Word like the navigation pane and Track Changes. Learning them will save their time and yours, especially for longer documents. Many students do not know how to use page breaks, section breaks, or paragraph settings like "keep lines together." Small caps may be needed in engineering.

Teaching tip: The rough draft can be written over several weeks while you begin to talk about rewriting, from the following chapters. Writing the first section will take the longest. Allow time for rewriting one division, at least at the level of organization, before writing a second.

Notes

1 See section 12.1.
2 This is especially useful when working on a long work like a thesis, as the section headers can be used to automatically generate a Table of Contents. If there are blank lines in the navigation pane you added a line return at the end of a heading. Click on the blank line in the navigation pane to go to it, then click on Normal style instead of Heading.
3 This happened to a student at Cornell while I was there.
4 As one of my students helpfully noted, for beginning writers this is like baby poo, small and poorly formed. Analogies about the importance of writing daily to prevent various problems I will leave to your imagination.

References

PULLMAN, P. 2017. *Daemon voices: Essays on storytelling*, Oxford, David Fickling. Quoted at https://writerswrite.co.za/philip-pullman-on-writers-block-2.
SAMUELSON, A. 1984. *With Hemingway: A year in Key West and Cuba*, New York, Random House.

Chapter 21

Rewriting for organization

21.1 Rewriting in stages

As I have said from the beginning, rewriting is the most important and most time-consuming part of the writing process, even for the best writers. Ernest Hemingway rewrote the final words of *A Farewell to Arms* at least 39 times before he was satisfied. Your work may not need quite that many revisions, but it probably will not have the same audience or impact either. Nevertheless, as the 19th-century German physician and physicist Hermann von Helmholtz testified, "The writing aspect of scientific research is exhausting . . . I have rewritten many parts of papers four to six times, restructuring the entire organization, before I finally became satisfied" (in Alley, 1996: 228).

For experienced writers, much rewriting will occur during the process of writing, as they delete and rewrite whole sentences or even paragraphs. They will not follow a discrete series of steps but will change things on multiple levels at once. Nevertheless, they will try to make large changes before small, and will go through the whole text many times, as I am suggesting you do.

> The 7Cs of Change is a scaffolding, not a method to follow exactly forever. Nonetheless, it is a helpful device for beginning writers.

Do not feel that you need to slavishly follow the order suggested. As with other ideas in this book, it is a scaffolding, a way of laying out what needs to be done, rather than a method to follow exactly. On the other hand, there is a logic to the order presented. It makes sense to change overall organization before worrying about specific word choice. Therefore, for now I suggest completing the steps in order, reading about one and practicing it before moving on to the next. That will help you see what you do well and what you need to work on. There are online resources for each topic, once you know what you need more help with.

As your writing improves, you will develop your own method. The method that works for me may not work for you. You may prefer to write a rough draft of a whole division or article before going back to rewrite or may rewrite one section at a time. Some might go back and forth from one section to another as thoughts come, while others stick mostly to one section, jotting down ideas relating to other sections as they occur. Some will prefer to begin writing each day without reviewing what they wrote the day

Table 21.1 The 7Cs of Change, highlighting those covered in this chapter

Type of change	*7Cs*	*Purpose*	*Level*	*Scope*
Rewriting	1 **Coherence**	**Organization**	Paragraph	Section
	2 **Conciseness**		Sentence	
	3 Connection	Clarification	Sentence	
	4 Connotation		Words	
Revising	5 Consistency	Finalization	Words	Whole paper
	6 Correctness		Format	
	7 Collaboration		All	

before, while others will reread and rewrite the previous day's work before continuing. Nevertheless, however you do it, you will have to read and reread, write and rewrite many times, especially at graduate level and beyond.

In this chapter I will talk about coherence and conciseness, rewriting to improve the organization of a paper, as shown in Table 21.1. Experienced writers tend to write more than necessary and may prefer to tackle conciseness first, but novice authors generally find it more beneficial to check first for coherence. Since both deal with the overall structure of the argument, as you gain experience you will probably find yourself doing both at the same time.

21.2 Coherence

Let me begin this section with an analogy: academic writing is like a tailor-made suit. When complete, you want it to completely cover the subject, without feeling tight, but without excess material. To accomplish this, before you begin cutting you need to plan, take measurements, and lay out the pattern to make sure you have enough material. You must make sure that once you cut out the pieces and sew it together, the pattern will match. You don't want the stripes running vertically on one half and diagonally on the other, and the stripes on the two sides should be mirror images. In order to have strong coherence at the end, you need to work it in from the beginning, as it is difficult if not impossible to add it later.

Thus the first stage of rewriting is to check the coherence, making sure you have enough material to cover your subject and checking the argument structure. This is one of the most difficult practices to teach, because of the various ways arguments can be structured, but there are a few general principles.

> Arguments usually move from known to new information, following an organizational sequence predictable to readers.

An important general rule of academic writing, and indeed all communication, is to put known information first. This is the reason past research, which your audience should be familiar with, is described before showing how your new work adds to it. Similarly, ideas you have already mentioned in a document are generally placed before new ideas in a paragraph or sentence.

Recall that a claim may be stated either before or after the support, or both. For example, in engineering the research goal often states the main claim as the contribution of the research. This is supported throughout the document, then restated in a slightly different form in the conclusion.

It is also possible to proceed either from a general theory to specific examples or from specific examples to a more general theory or model. The former is far more common in the academic literature, since new models are relatively rare.[1]

The argument often also follows a recognizable organizational scheme (section 9.5). This can be chronological or the operational sequence of your design, but the order should be clear and predictable to your audience.

Since there are so many factors to consider in terms of how to organize a document, it is impossible to prescribe a general order. Choose an order that allows you to present your argument in the clearest way possible. This frequently means rewriting it several times until you find an order that works well.

When revising, begin with the largest units and move to the smaller.

Any document needs to be coherent at multiple levels of organization, from sections to paragraphs to sentences, and needs to be rewritten in that order. In longer works such as a thesis or book you may choose to reorder the chapters after you begin writing. Organization of sections should be straightforward for a journal article, so if your prewriting was effective you will mostly consider the order of paragraphs and sentences within each section.

The first step, then, is to reevaluate the order of the paragraphs. First remind yourself what claim or claims you are arguing in that section and what components you need to include. Then ask, does each paragraph have its own purpose and topic, and does the order make the argument easy to follow? You may need to reorder the paragraphs, splitting ones that do not have a single main topic or combining ones that contain similar ideas. As a first-time author you will probably need to add more material at this stage, as you find gaps in the argument.

The second step is to check the order of the sentences, striving for logical flow of thought within each paragraph. In the process, you may find some sentences that should be moved to a different paragraph, which will necessitate rewriting both paragraphs.

In summary, ask yourself these questions about every paragraph and sentence:

- Does each have a distinct purpose and topic?
- Does the order make it easy to follow the argument?

Exercise 21.1 Evaluating the coherence of your writing

(1) Beginning with the general claim appropriate to whatever you are writing, write a related specific claim.

Hint: For example, for the Introduction of a research article, instead of the first general claim you might say, "[My research goal] is important because [reasons] and it expands on [related research]."

(2) Next, evaluate the coherence of the argument you have written to support that claim. What is the supporting evidence? What is the purpose of each paragraph, each sentence? Is the order logical, based both on some organizational scheme and the novelty of the information?

Hint: For example, does your Introduction make both main points—importance and need? Are they in the right locations and following the organization expected in your field? Are they supported by an appropriate number and type of citations?

21.3 Conciseness

Conciseness is an essential feature of scientific writing. Not only does it make it easier for the reader; focusing on the main point also sharpens the understanding of the writer. As Thomas Jefferson wrote,

> This operation is doubly useful, insomuch as it obliges the student to seek out the pith of the case, and habituates him to a condensation of thought, and to an acquisition of the most valuable of all talents, that of never using two words where one will do.
>
> (in Mayo, 1942: 17)

Returning to our example of the tailored suit, you need to begin with more material than is needed in the final product. It is easy to cut material out, but it is much harder to stitch something in later without destroying the unity of the garment. In the same way, it is far easier to begin with an excess of material as you brainstorm and write your rough draft, but then the process of cutting begins.

You need to eliminate three things from your writing: off-topic ideas, repetition, and wordiness. Hopefully between the time you did your brainstorming and the time you wrote your first draft, you eliminated most of the excess material. Now that you have a first draft, it is time to repeat the process. It is easier to do this in two stages, before and after you check the order for coherence.

Before you begin to reorganize for coherence, read through what you have written and eliminate anything that is off-topic, does not address the goal of this writing task, or is not necessary to your argument. Be ruthless. Scientific writing is concise. If what you have written is not relevant to the question addressed in this paper, even if it refers to an important study in the field, cut it.[2]

Eliminate anything that is off-topic before checking for coherence, and anything repetitive or wordy afterward.

The second step is to look for repetition. If you have written similar things in two sentences, try to eliminate one or merge the two. Often this will involve trying several wordings before you settle on one that is as concise as possible without loss of meaning. Sometimes rather than combining two sentences it is easier to set both aside and rewrite the same thought from scratch.

The third step is to eliminate wordiness. As Jefferson said, "never us[e] two words where one will do." For example, instead of saying, "a sufficient number of tests were done," you could say, "sufficient tests were done." Phrases can sometimes be replaced by single words, for example replacing "in the vicinity of" with "near." Sentences can sometimes be reduced by simplifying the structure, for example "there are two things that could be improved" is identical to "two things could be improved."

> How concise your writing needs to be depends on the style of writing and the expectations of the readers.

How much you cut depends in part on the style of writing. Far more will be cut from condensed style than extended. In condensed style, any topic sentences used to organize your writing should be removed at this stage, sentences combined using relative clauses and other structures, and every word weighed and eliminated if not absolutely necessary. In expanded style, a certain amount of word repetition is expected as a way of linking sentences and paragraphs, but thoughts are not repeated.

Exercise 21.2 Evaluating the conciseness of your writing

(1) Evaluate the conciseness of your rough draft, making whatever changes are necessary.

- Does every paragraph and sentence relate to my goal?
- Is every paragraph and sentence necessary to my argument?
- Is there any unnecessary repetition of ideas?
- Could any words be eliminated without loss of meaning?

Hint: For most academic writing, the goal is stated in the research goal, but also consider the larger overall goals discussed in Chapters 11–15.

Hint: As always, begin with larger units and move to smaller.

Evaluate your understanding

(1) How can you evaluate the coherence of your writing?
(2) Why should you always begin with large changes like the order of paragraphs or sentences, before working on sentence structure, grammar, and word choice?
(3) What should you eliminate as you revise for conciseness?
(4) How do you know when your writing is concise enough?

Homework

(1) Rewrite your draft, concentrating on making it coherent and concise.

Hint: Save a copy before you start cutting and give your cut version a new filename. That way if you want to go back to restore something that was cut, you will be able to find it.

Hint: If possible, show this to your instructor or advisor and get comments on it before moving to the next chapter or the next stage of rewriting. Make sure he or she knows it is a rough draft—that you only want to check that all necessary information is present and well organized.

Teaching tip: If possible, work closely with the students' advisors so they do not do too much editing, but allow the students to do as much as they can in each step of the process.

Teaching tip: Reverse outline is also a good procedure to use at this stage to evaluate the organization. Since it is a standard technique, I have not included it here.

Supplemental material:

Class exercise 21.1 Practice: coherence 1 (with answer key)

Class exercise 21.2 Practice: coherence 2

Class exercise 21.3 Practice: conciseness 1 (with answer key)

Class exercise 21.4 Practice: conciseness 2

Notes

1 Most normal science is based on deductive argumentation (general to specific), while inductive arguments (specific to general) can lead to major paradigm shifts (see Class exercise 21.3).
2 There are, of course, exceptions to any rule. Sometimes it is expedient to mention at least one publication of someone who is likely to be a reviewer of your article, even if that research is marginal to your work, but that is a nuance you will figure out as you write more.

References

ALLEY, M. 1996. *The craft of scientific writing*, New York, Springer.

MAYO, B. 1942. *Jefferson himself: The personal narrative of a many-sided American*, Charlottesville, VA, The University Press of Virginia.

Chapter 22

Rewriting for clarification

22.1 Clear, concise, precise

During the next step of rewriting your goal is to make your arguments as clear as possible. The easiest way to do this is to restate your points, using extra words to ensure your meaning cannot be mistaken. However, writing in science and engineering must also be concise, as we found in the last chapter.[1] Therefore you must write precisely, making sure every detail, every word, is clear and accurate. This makes sentence structure and word choice extremely important.

> Any time you move one sentence you will have to rewrite the whole paragraph, so reorganize before worrying about word choice.

As the novelist Philip Roth put in the mouth of one of his characters,

> "I turn sentences around. That's my life. I write a sentence and I turn it around. Then I look at it and I turn it around again. Then I have lunch. Then I come back in and write another sentence. Then I have tea and turn the new sentence around. Then I read the two sentences over and turn them both around. Then I lie down on my sofa and think. Then I get up and throw them out and start from the beginning."
>
> (Roth, 2011)

Almost every change within a sentence, whether a choice of one word over another or even changing the form of a word, for example from participle to infinitive, requires changes in the structure of that sentence and surrounding sentences. That is what Roth means by turning sentences around. Obviously, it is pointless to try to find exactly the right words until your basic argument structure is clear. Otherwise, it would be like polishing a sculpture before you finish carving. Sometimes it is impossible to avoid, when you realize you have left out some important element or need to cut to reduce your word count, but try to avoid it as much as possible.

In this chapter I will talk about two ways of clarifying your argument: using *connectors* to link the argument together and *connotation* to make your claims easier to support (Table 22.1). I want to emphasize again that, at least in science and engineering, the purpose at this stage is not to show off your proficiency in English but to strengthen the

Table 22.1 The 7Cs of Change, highlighting those covered in this chapter

Type of change	*7Cs*	*Purpose*	*Level*	*Scope*
Rewriting	1 Coherence	Organization	Paragraph	Section
	2 Conciseness		Sentence	
	3 Connection	**Clarification**	Sentence	
	4 Connotation		Words	
Revising	5 Consistency	Finalization	Words	Whole paper
	6 Correctness		Format	
	7 Collaboration		All	

argument. Neophyte wordsmiths may opt for grandiose terminology, perchance supposing to impress their audience, notwithstanding that the only tangible consequence is obfuscation of their contention. Or perhaps I should say, using big words does not make sense if it only makes your writing harder to understand.

This chapter is longer than others because details are extremely important at this point. Additional online materials will help those using English as a lingua franca. Clarifying arguments and shades of meaning is one of the hardest tasks for any writer, let alone those writing in a second language.

Clarifying arguments is one of the hardest but most important tasks for any writer, involving many details. Additional materials are available online.

22.2 Connection

Returning to the analogy of a tailored suit, once you have laid out the pattern, cut the pieces, and pinned them to one another, it is time to stitch them together. Without that connection, the suit is not finished. Similarly, you may have done wonderful research and written each individual result well, but if it is only roughly pinned together no one will appreciate the quality of the material.

Connectors are used *to improve continuity of thought and thus clarify an argument.* First I will show how to better connect your work with previous research in the field, then how to improve connections within your writing.

Connectors clarify your work, either by connecting your work with prior work or showing the logic of your argument.

22.2.1 Connection to past research

Showing how your research connects with previous research can be done using what I will call contribution markers, achievement markers, and application markers.

These are not grammatical categories, but rhetorical devices, using a word for a specific communicative purpose. All three are found in science, while the first two are more common in engineering. Common categories of each marker type are shown in Table 22.2 and discussed below. Examples of each category can be found in the online supplemental materials (Supplemental Tables 22.1–3) or by using the synonym and thesaurus functions in Word.

Contribution markers

The first claim in a research article could be restated as *this work connects with and makes an important contribution to the overall body of knowledge in the field.* Beyond citation of previous work, the basic method for supporting this claim is contribution markers.[2]

> Contribution markers are verbs that show how the research supports, challenges, or modifies previous work, or solves a problem.

Contribution markers are *verbs or verb phrases* that show how your work supports, challenges, modifies, or solves problems with previous work. Since science proposes explanations of data and phenomena, the first three categories predominate. Engineering commonly modifies designs or solves a problem. The four categories are illustrated by the following examples:

- Our data corroborate . . .
- These findings contradict . . .
- Our study expands on . . .
- This paper has resolved the issue . . .

Contribution connectors appear throughout a document. Present tense is frequently used in the first three divisions, highlighting how the current paper is contributing, as illustrated by the first three examples. When reiterated in the Conclusion, the present perfect is common, as a final summary of what has been contributed by the paper, as illustrated in the last.

Table 22.2 Connection marker types and categories

Marker type	*Categories*	
Connection	Support	Modify
	Challenge	Solve
Achievement	Excellence	Efficiency
	Importance	Cost efficiency
	Novelty	Feasibility
	Perfection	Strength
	Accuracy	Clarity
	Utility	Emotive
Application	Utilize	Beneficial
	Apply	Potential
	Facilitate	

Achievement markers

While connection with past work is important, you must also *highlight your achievement*; what makes your work unique. Achievement markers are *adjectives or adverbs* that describe your research or results in a positive way. Some give a general impression of excellence, others draw attention to specifics such as low cost or efficiency, and yet others evoke positive emotion.

Achievement markers are adjectives or adverbs that highlight your contribution and why it is worth publishing.

Achievement markers are far more common in engineering than contribution markers, highlighting specific positive characteristics of the new design. Perhaps another analogy will help explain this. Contribution markers are like the main ingredient in a recipe, whereas achievement markers are like the spices. The main ingredients in your recipe may be almost the same as everyone else's, but the seasoning is what makes one better than another. Your contribution may be similar to others, but you want to highlight what makes it special, so sprinkle achievement markers liberally throughout your writing. On the other hand, like spices, make sure that they are not too concentrated in one place or they will ruin the dish.

Obviously, achievement markers are not a replacement for good research, but can help strengthen the claim of your accomplishment, as shown in the following examples. In each pair, the first is a basic statement of the contribution while the second shows how adding markers highlights the achievement.

- The data challenge the interpretation proposed previously.
- The extensive data clearly and convincingly challenge the interpretation proposed previously.

- Our prototype demonstrates that the fabrication method works.
- The unprecedented efficiency of our prototype demonstrates the feasibility of this simple, low-cost fabrication method.

- This study presents a solution to the problem of circuit density outlined by Chen et al.
- This study presents a novel, cost-effective, flexible solution to the longstanding problem of circuit density outlined by Chen et al. more than ten years ago.

Note in the first example that both adjectives and adverbs can be used, in the second that the sentence structure may need to be modified, and in the third that phrases can also be used to highlight the achievement. In addition to the general categories of achievement listed in the table, many others are field-specific. For example, there may be a value to something being high-capacity, compact, or rapid-cycling.

Application markers

Engineering seeks a solution to a current problem, so the application of the research is clear from the beginning. Although science seeks answers to basic questions, there are

> Your work may also have possible applications that have not yet been proven. These can be stated using application markers, with modals to indicate potential.

usually possible future applications. Application markers *draw attention to potential applications or their benefit.*[3]

When discussing possible future applications, it is important to use either modals or words such as "potential" that indicate the results are possible rather than actualized, as in the following examples.

- This approach can be applied . . .
- This research could lead to . . .
- This technique might be beneficial in developing . . .
- This information holds potential for resolving the question . . .
- Based on these results, it should be possible to . . .

Exercise 22.1 Highlighting your contribution

(1) Identify contribution, achievement, and application markers in your exemplars. How many of each are there? Which components are they associated with?
(2) What sentences or words could you add to tie your work better with both past and future research in your field while at the same time highlighting the importance of your contribution?

Teaching tip: Have students share their findings, then discuss why those might be common in their field.

22.2.2 Connection within your article

If a paper is hard to follow, readers may never reach the end of it. Perhaps another illustration would make the point more convincingly. Potholes and bumps in the road force a driver to slow down. Sudden unexpected turns may cause a driver to go off the road or get lost. Moreover, if a road is washed out, you can see the other side, but it is impossible to get there. Similarly, if you want readers to follow your article to the end, you must make it easy for them. There are many ways to improve the smoothness and continuity of your article, including continuity connectors, logical connectors, and time sequence connectors. The first is more common in extended style, while the other two are found in both.

> In extended style, sentences are often linked by repetition of words, synonyms, or relative pronouns.

Continuity connectors

Continuity connectors *link sentences* or *phrases* by repeating an idea, either by repetition of a word or use of a synonym, pronoun, or relative pronoun. Continuity connectors are less common in condensed style, but some are present in all writing.

The simplest way to link sentences is to repeat the same word in subsequent sentences. As shown by the example below, this is an effective way of stressing a keyword when it is first introduced or when the second sentence defines that new word.

- The process of science involves making logical inferences. An inference is a conclusion, explanation, or judgment based on the evidence.[4]

Instead of repeating the word, a synonym, pronoun, or relative pronoun can be used. Synonyms are less common in science because of the desire for consistency of terms, but could be used, as shown in the following example. Note that whereas in the first example the end of one sentence was immediately repeated at the beginning of the next, in this example the connection is between the initial words of each sentence.

- The difference between evidence and inference is essential to the question of origins, as indeed it is in every area of science, so we must learn to clearly differentiate the two. The distinction between the two will be a major emphasis of chapters three through six.[5]

To be more concise, a relative pronoun can be used. Relative pronouns can link either sentences or clauses, as shown in the following two examples.

- The difference between evidence and inference is essential to the question of origins, as indeed it is in every area of science, so we must learn to clearly differentiate the two. This will be a major emphasis of chapters three through six.
- Each scientist is working from the perspective of one particular theory, which affects both data collection and interpretation.

Less commonly, personal pronouns may be used when a subject is anthropomorphized, given a human characteristic such as the ability to cause an effect, as illustrated in the following example.

- Models may be explicitly stated or implicitly held, but they direct our research in the same way our worldview and philosophy subtly shape our thoughts. They have been shown to affect both the choice of what data to collect and how the data are interpreted.

The least common continuity connector in science and engineering is the semicolon, used to join two very closely related independent phrases. One phrase may also include a relative pronoun, as in the following example. The author could have chosen to use a comma and "which" instead of a semicolon and "that," but used the semicolon for emphasis.

- Usually, however, it is relatively easy to list all possible classes of alternatives; that is the case with the question of origins.

Logical connectors

Next we will turn to logical connectors: words or phrases that *alert readers to the relationship between different clauses or sentences*. Although many methods have been used to group logical connectors, I propose eight categories, as shown in Table 22.3. The last

Table 22.3 Types of logical connectors

Type	*Connection*	*Examples*
Addition	A and B	additionally, furthermore, moreover
Contrast	A vs. B	however, on the contrary, whereas
Result	A causes B	therefore, consequently, thus
Cause	B caused by A	because, since, as a result of
Unexpected result	A does not cause B	however, nevertheless, even so
Unexpected cause	B not caused by A	although, in spite of, regardless of
Clarification	A restated as B	in other words, that is, i.e.
Illustration	A exemplified by B	for example, for instance, e.g.

Logical connectors help readers follow the argument.

two types are rare in natural science and engineering because of the value placed on conciseness, but common in social science. Several words fit in more than one type, depending on the context. Further examples of each type and how to change sentences to incorporate them can be found in the online supplemental material.

Most types are obvious, but two require explanation: unexpected result and unexpected cause. Both report results that are unexpected, surprising, or contrary to predictions. Commonly grouped together as "unexpected" or "adversarial," I have separated the two and chosen names that reflect their relationship to result and cause connectors. More detail on the logical connectors, and practice exercises useful for those using English as a lingua franca, can be found in the supplemental materials online (Supplemental Tables 22.4–8).

Addition connectors take the place of the conjunction "and," but add emphasis to the phrase after the connector.

Contrast connectors take the place of the conjunction "but," and vary the strength of the contrast implied.

ADDITION

The first type of logical connector, addition, is a way of joining two similar ideas. Unlike the simple coordinating conjunction "and," *these connectors add different degrees of emphasis to the second idea*. Words or phrases with the idea of "addition" or "more" indicate the second idea is more important. Phrases like "and in fact" tend to introduce something unexpected. Numbering (firstly, secondly, . . .) makes it easier for readers to follow a series of steps in an argument.[6]

CONTRAST

The second type of logical connector alerts readers to a *contrast between the preceding idea and the following idea.* The coordinating conjunction "but" is the simplest way to express contrast.[7] Some like "conversely" or "on the contrary" imply a stronger contrast than "however."

Contrast connectors also provide another way of challenging previous work:

- Contrary to what was reported previously, we observed that . . .

RESULT AND CAUSE

Result and cause connectors *both express the same causal relationship, but in the opposite order.* Result connectors take the form *A causes B*, whereas causal connectors have the basic form *B is caused by A*. As with addition and contrast, it is possible to express result and cause using coordinating conjunctions, in this case "so" and "for," respectively.

> **Result and cause connectors replace "so" and "for," respectively, highlighting the causal relationship.**

In most cases, a relationship can be expressed either as a result or a cause, with the phrases inverted, as in the following:

- The equipment was not properly calibrated, therefore the data were inaccurate.[8] (Result)
- The data were inaccurate because the equipment was not properly calibrated. (Cause)

Causal connectors always come at the beginning of a dependent clause or noun phrase, never an independent clause, but in English it is possible to invert the sentence so that the dependent clause comes first, separated by a comma. Thus, although the basic form in terms of the logical structure is *B is caused by A*, the *actual* order in the sentence may be inverted, as shown below.

> **The clauses of sentences using causal connectors can be reversed, putting the causal connector at the beginning to alert the reader.**

- Because the equipment was not properly calibrated, the data were inaccurate. (Cause)

Note that each clause is identical to the previous example, but the order is reversed. On the contrary, if the order of the clauses were reversed for the result connector, it would make no sense. You could not say, "Therefore the data were inaccurate, the equipment was not properly calibrated."

This ability to invert the order is quite useful. Putting the causal connector first alerts the reader to the causal relationship at the beginning of the sentence. By contrast, with either result connectors or the basic order for causal connectors the reader is not alerted to the causal relationship until the middle of the sentence, making it difficult to understand if the sentence is long and complex.

Beyond connectors, there are other ways to represent causal relationships. Verbs such as cause, affect, or lead to, and nouns like cause, effect, or result, can also be used. The sheer number of ways of expressing the same concept indicates its importance.

> **Unexpected result and unexpected cause connectors show results that are the opposite of what might have been expected, and can replace the conjunction "yet."**

> **Sometimes it is difficult to separate unexpected results from contrast, so both sets of connectors may be used, but with slightly different connotations.**

UNEXPECTED RESULT AND UNEXPECTED CAUSE

These two categories are named for their similarity to result and cause connectors, except that something *did not cause the expected result.* Thus they express a surprising relationship between a situation and an unexpected result. Unexpected result can be expressed by a coordinating conjunction "yet," but there is no conjunction to express unexpected cause.

Sometimes it is difficult to clearly separate unexpected results from contrast. For example, comparing an odd case with what usually happens could be considered a contrast of two categories or an unexpected result. Thus, "however" can be used for either. As illustrated in the sentences below, choosing a word from either category will in many cases yield a sentence that makes sense, although the emphasis is different and some sentences may need a little rewording.

- Outliers are usually ignored, (but/yet) this one led to a new discovery.
- Outliers are usually ignored. (Conversely/However), this one led to a new discovery.
- Outliers are usually ignored, (while/even though) this one led to a new discovery.
- Outliers are usually ignored, (unlike this one that led/despite the fact that this one led) to a new discovery.

As with result and cause, the number of ways that contrasting or unexpected events can be stated indicates their importance. Certainly this is the case in science and engineering, where the unexpected or anomalous observation may lead to a new and deeper understanding of a phenomenon. Consequently, it is important to highlight such data for your readers using appropriate connectors.

CLARIFICATION AND ILLUSTRATION

The final two types of logical connectors are *used to restate or expand on a statement.* These are far more common in the humanities and social sciences than in natural science and engineering, but are included here for completeness. Although the connectors "that is" and "for instance" can begin an independent clause, the associated abbreviations i.e. (id est) and e.g. (exempli gratia) are rare in sentence initial position. When using either abbreviation, standard American usage (but not British) requires a comma after it.

Clarification and illustration connectors restate or expand on a statement, and are more common in social science.

Time sequence connectors

Time sequence connectors denote a sequence of events, either in the current research or the history of the field.

Your argument can also be strengthened by proper use of time sequence connectors. These may *denote the sequence of events within the course of the current research or change in thought over time in the field.*[9] They may also be used when discussing future possible applications or developments of your idea. Examples may be found in Supplemental Table 22.9. Ordinal numbers (first, second, third) can be used for either logical or time sequence.

Although time sequence connectors can be used either at the beginning or in the middle of a sentence, it is far more common in concise scientific writing to place them at the beginning for emphasis. However, when actual time is recorded, rather than sequence, it often falls in the middle of the sentence. This is illustrated in the following examples, the first two from (Hu et al., 2013), the latter two from (Kleinheinz et al., 2009).

- Initially, the DO obtains a private and a public key through a trusted key distribution center.
- Then, it sends the signatures and the data to the SP which constructs the necessary data structures for efficient query processing.
- By the time the eight-hour samples were collected, the difference was no longer significant.
- Samples were analyzed within four hours of collection.

Exercise 22.2 Adding connectors to strengthen your argument

(1) Look in your exemplars for continuity, logical, and time sequence connectors. Circle each, then list the most frequent types and words employed. Note where and how each are used.

Hint: Are all types equally common in all divisions or with all components?

(2) How can you improve your use of connectors to clarify and strengthen your argument?

Hint: Organizational strategies (Chapter 9) are also important, of course.

22.3 Connotation

We now turn to the fourth of the 7Cs, connotation. Connotation refers to the *fine shades of meaning conveyed by the choice of words or modifiers*. It is obviously pointless to worry about fine shades of meaning until the basic meaning is well established. If the basic structure of the paper is good, this stage will be relatively easy, a mere tweaking of the words. If the basic structure still needs realigning, changing a few words will not make any noticeable improvement.

Returning to our analogy of a tailor-made suit, connotation is like adding the fasteners—the buttons, clasp, and zipper. Their absence or failure would quickly be noted by others, so it is essential that they be as high quality as the rest of the suit. Nevertheless, they cannot make up for lack of quality in the overall process of selecting the material and cutting and stitching it together.

> In science and engineering, connotation is often used to convey fine shades of meaning related to the strength of claims.

Connotation is important in all types of writing but serves different purposes and therefore employs different techniques. In literary writing connotation makes the material more interesting, conveying details. In academic writing it is used primarily to modify claims, making them easier to support. Thus, in this section we will first discuss the general concept of making supportable claims, followed by the proper use of statistical terms when making claims. Finally, we will consider how to modify claims.

22.3.1 Making supportable claims

> You want to make the strongest claim you can support, without antagonizing readers who may disagree. Qualifiers allow you to fulfill both conditions.

You want to claim that you have a better explanation of a question or a better solution to a problem than any that has yet been presented. Yet among your reviewers and readers are those who have presented what you claim to be bettering. Aye, there's the rub. How can you support your claim without antagonizing your readers? Therein lies the reason for using qualifiers.

Qualifiers allow you to hedge your bets; to say what you think is the most likely explanation without putting all your eggs in one basket. They allow you to claim your solution is the best in a certain situation, if not every situation. It is hard to defend a claim that your answer or your solution is the absolute best, so qualifiers make a claim easier to defend by limiting its scope. Sometimes as you approach the end of your writing you will find that you cannot fully support your initial claim. Instead of rewriting the whole paper, you can modify the claim to make it more supportable.

Let's consider in general terms how this might work by returning to the list of basic claims introduced in Chapter 3, reproduced here as Figure 22.1. I suggested that research articles make several implicit claims.

> No one advertises their product as second best; on the other hand, you cannot directly say similar products are bad. The same is true in research.

The first claim is like advertising a new product. No advertiser says their product is second best, or almost as good as some other. On the other hand, if you denigrate similar products you will suddenly find yourself in the middle of a nasty and costly lawsuit with your competitor. Similarly, if you claim your work is the best and others are junk, some of your readers (who *know* their work is the best) may write a critique of your work without even reading it.

For the first part of this claim, although you view your own work as important, you need to convince others that it is worthwhile. (If you do not think your work is important, I suggest you find a new field.) As Benjamin Franklin said, "If you would not be forgotten as soon as you are dead and rotten, either write things worth reading, or do things worth the writing" (Franklin, 2013: 78). Stating importance can be done without criticism of others. Start with something everyone can agree on, so they are willing to read further.

The second part of the first claim is your research goal. This may be uncontroversial in the natural sciences, where the research gap often lies at the border of current research, unless it challenges previous findings. In engineering, with its emphasis on solving a current problem, the field is much more crowded, and often your proposed solution is in direct competition with a current solution. If so, you will need to be more careful about how you state the need for your research.

Implicit claims in science and engineering research articles

IMRD (science)

1. You have identified an important unanswered question
2. You have gathered data in a valid, reliable way
3. You have a good explanation for your data

IPTC (engineering)

1. You have identified an important unresolved problem
2. You have designed a workable solution to the problem
3. Your solution is better than existing solutions

Figure 22.1 Implicit claims in science and engineering research articles.

The second claim is far different in science and engineering. In science, behind the careful statement of the methods used is the implicit claim that the data are reliable, the methods themselves being secondary. On the other hand, the description of your product or process in engineering substantiates your central claim to have developed a better solution. Nevertheless, these sections are descriptive, with little if any inference. Others may dispute whether you used the best method, but they cannot dispute what you did. If you wrote this section as you went and revised it to make sure it is concise, coherent, and the connections between different parts are clear, you should not have to worry about connotation for this claim.

Connotation is most important for the third claim. In IPTC, it may be necessary to qualify your claim by saying your solution is only slightly better, or better in certain situations. In IMRD you are claiming that your inferences, your interpretation of the data and explanation of the causal relationship, are correct. As you consider alternative interpretations, qualifiers are used to indicate that this is the best supported explanation, but not the only one.

Keep in mind that overall the article claims that you have made an important contribution, however small. With few exceptions the contribution of any single article is very small, so this is not just a matter of humility. Whether it turns out to be more important depends largely on how well it stands the test of time and how many people cite it, so the claims and support must be clear and not antagonize others.

22.3.2 Proper use of statistical terms: significant

Statistics are more important in science than engineering. Regardless of your field, I suggest reading this and the following subsection to avoid misusing terms that have a specific technical meaning. Full treatment could not be attempted apart from a presentation of the statistical tests themselves and the assumptions behind them, so I will simply highlight a few common mistakes.

In general, when the term "statistics" is used in the natural sciences and engineering it refers not to data values themselves, as it often does in the social sciences (e.g. "If we look at the statistics for the unemployment rate over time . . ."), but to inferential statistics. Statistical tests ascertain how likely it is that parameters calculated from our sample data represent true characteristics of a larger set we are interested in.

> "Significant" should only be used if a statistical test of significance has been done to show the result is not likely to be the result of random variation in measurement.

One of the biggest mistakes made by those who do not use statistics regularly is misuse of the word "significant." In statistics, this has a specific technical meaning, that the result obtained is highly unlikely to have been obtained by chance variation alone. *If no statistical test has been performed, the word "significant" should not be used in scientific or technical writing.*

If a statistical test has been done, the significance level should be specified, often 0.05 or 0.01, meaning that a result that extreme would be expected to happen only one time out of 20 or one time out of 100, respectively. Note that there should always be a number to the left of a decimal point, even if that number is zero. Keep in mind too that some tests that are significant do not indicate a real difference from expectations. One time out of every 20 times you do a certain test you would expect the results to be significant at the 0.05 level. Thus, claims based on statistics must be written carefully. A comparison of means cannot show that the two means are identical, but rather that they are not different at a certain level of significance.

One of the most common errors made by those unfamiliar with statistics is to say something like, "A was larger than B, although the results of the test were not significant." *If the results were not significant, you cannot say A was larger than B*, even though the value you measured was larger. Not significant means that if you measured again you might find that the measured value of B was larger than A, because the difference between the two could easily have been caused by sampling error. If you believe that A is truly larger than B, even though the results do not show significance, you will have to be very cautious about qualifying your claims and justifying why you believe it, perhaps because of multiple tests that cannot be combined statistically but all show the same trend or triangulation of the data with other independent measures.

22.3.3 Proper use of statistical terms: causation vs. correlation

Much of the work in science, and some in engineering, seeks to identify the cause of a phenomenon. Sometimes, however, it is difficult to separate causation from correlation. In a scientific sense, *cause is the capacity of one variable to influence another in a predictable way*, either by bringing the second into existence or modifying it in some way. *Correlation is a relationship between two variables such that both change together in some predictable way*, either because of their action on one another or some other factor that affects both. It is essential both to clarify what you think is going on and to state the claim properly.

Causation could be symbolized as:

A ➔ B.

That is, A causes some change in B, or as A changes, there will be a resultant change in B. One example would be if A is the number of times a wire is coiled around the magnetic core of an electromagnet, and B is the strength of the electromagnet. As A increases, B will also increase at a predictable rate, as a direct result of the number of coils. Of course, this does not answer the question of *why* that is true—what the underlying mechanism is—but it will suffice for now for the sake of illustration.

Correlation, on the other hand, simply means that there is some unknown relationship between A and B. As A changes, there will be a concomitant change in B, but the causal relationship could be any of the following:

A ➔ B, B ➔ A, or C ➔ both A and B.

Many times, work on causality will begin by noticing a correlation. Causation can only be proven, however, by eliminating all other possible causes. It is also possible that something may be *a* cause, but not *the only* cause. Or perhaps you know it is correlated, and have not yet proven causation, but want to indicate that you think it is likely. So what is it that you want to claim?

- A caused B
- A partially caused B
- A may have caused B
- A is highly correlated with B
- A is slightly correlated with B
- A may be correlated with B

"Correlated" should only be used if a statistical test of correlation has been done.

There are many ways to express both causality and correlation (see Supplemental Table 22.10). The relationship between the variables can be expressed either as cause or result, just as relationships between ideas in your writing can be equally expressed by cause or result connectors. *The term "correlated" should only be used if you have done a statistical test of correlation and that test is significant.* If you have not done that test, use another expression.

Of course, even if the statistical correlation is very high, it is possible that there is no relationship between the two at all. For the ten-year period from 2000 to 2009, there was a 98.51% correlation between the number of computer science doctorates awarded in the US and the annual revenue generated by arcades. Did the computer scientists' work increase the revenues (maybe by designing better video games?), or did the computer scientists get doctorates because they spent so much money in the arcades (looking for ideas on how to improve their own programming, perhaps?), or did some other social factor cause both to rise and fall at the same rate? Even more surprising, there was almost perfect correlation, 99.79%, between the US spending on science, space, and technology and suicides by hanging, strangulation, and suffocation for the period from 1999 to 2009. Perhaps both correlations were the result of a bored law student going on a "fishing expedition" in big data.[10]

Qualifiers allow you to modify the strength of your claim, making it easier to support.

22.3.4 Types and uses of qualifiers

Qualifiers allow you to modify the strength of a claim, making it easier to support. They can indicate how confident you are that your inference is correct. Some qualifiers cause your audience to view the data a certain way. As with connectors, I have grouped qualifiers according to their rhetorical purpose rather than their grammatical role. More examples of each category may be found in the supplemental online materials (Supplemental Tables 22.11–15).

Rarely can you be 100% convinced that you have established causation. Qualifiers modify the "truth value" of your statement, making it easier to support the claim. For example, it is easier to defend the statement "the presence of A appears to have strongly affected B" than it is to defend the statement "the presence of A strongly affected B." For the former, you only need to demonstrate probable cause—the likelihood that it *could* have had that effect. For the latter, you must eliminate all other possible causes, which is often difficult since it may not even be clear what all the other possible causes are.

There are several different ways of qualifying likelihood, including likelihood qualifiers, hedges, and delimiters. Frequency and similarity qualifiers can be used not only to modify claims about inferences or causality, but also claims about your evidence. Modals allow you to discuss possibilities or make suggestions. Verb connotations can also modify the strength of a claim. Depending on your purpose, it is often possible to combine several types of qualifiers in one sentence. This ability to combine qualifiers and use them in different contexts makes it difficult to assign them to a single role, so there will be some overlap between discussions of the different types.

Likelihood qualifiers

The simplest way to qualify a claim of causation is using what I will call likelihood qualifiers to specify *how likely you think your claim is to be correct.* There are many ways to express how certain you are about your claim, including degree of certainty, possibility, likelihood, and probability, which for most purposes are interchangeable.[11] Thus, the following expressions are nearly synonymous ways of making a very strong statement of your claim of causation, without making it absolute:

> Likelihood qualifiers indicate how likely you think your claim is to be correct, expressed as degree of certainty, possibility, likelihood, or probability.

- It is almost certain that A caused B.
- There is a very high probability that A caused B.
- There is a very strong likelihood that A caused B.
- There is a very strong possibility that A caused B.

Likelihood qualifiers are most likely to be found in the section(s) of the paper devoted to analysis, i.e. Testing or Discussion, where most inferences are presented.[12] Unlike some other qualifiers presented below, likelihood qualifiers can only be used with inferences or predictions, not with data, because once something has been observed, it has a probability of either 1 or 0—either it happened or it did not. Of course, if you can replace likelihood qualifiers with actual probability values, that is even better.

Frequency qualifiers

Frequency qualifiers are used to *draw attention to how frequently something occurs*, without giving an exact number. For example, they can be used to modify a claim of causality, as in

> Frequency qualifiers can be used either to modify claims or summarize data.

- A occasionally caused B.

Frequency qualifiers can also help summarize data. Although science has a reputation for being objective, data are always presented through the lens of the researcher. Scientists attempt to lead their readers to a certain conclusion without appearing to do so. As John Swales commented, "The art of the matter . . . lies in deceiving the reader into thinking that there is no rhetoric . . . and that the facts are indeed speaking for themselves" (Swales, 1990: 112).

Thus, one of the main uses for frequency qualifiers is to direct your readers to view the data in a particular way. Consider the difference between the following three statements:

- This occurred in 17% of the cases;
- This occurred rarely, in 17% of the cases;
- This occurred frequently, in 17% of the cases.

Each statement denotes the same thing, that the event occurred in 17% of the cases, but the connotation is quite different. The first is the basic statement, without qualifier. If the reader does not know the situation, it may not be clear whether 17% is higher or lower than expected. Adding the qualifier "rarely" changes the meaning, implying that this value is smaller than expected, while "frequently" has the opposite effect.

Whereas likelihood qualifiers are restricted to the probability of an explanation being correct or a future event happening, and therefore largely restricted to the analysis, frequency qualifiers often refer to past events and may be found in every division of an article. Consider the examples below, which are representative of ideas that might be found in each division, as summaries of previous work, methods, or aspects of the data:

- Introduction Generally it is thought . . .
- Process/Method Each time we . . .
- Testing/Results We never observed . . .
- Testing/Discussion Sometimes there were problems with . . .

Measurement qualifiers

Like frequency qualifiers, measurement qualifiers are used both to summarize data and to modify claims. Commonly, they *direct attention to how some measurement compared to another or to expectations*. Since this must be done frequently in an academic paper, it is useful to know different ways to express the same thought, to avoid repetition and make your paper more interesting to read. There are also small shades of difference in the meaning of the terms, which allows you to express your thoughts more precisely.

> Measurement qualifiers direct attention to how measurements compare to one another or to expectations. They can be used with either claims or data.

Words that express degree of similarity of measurements can be divided into several basic meanings (Supplemental Table 22.12). Some emphasize that two quantities are nearly equal. Others show that two quantities are close but not equal. Still others highlight the smallness of either a single value or the comparative difference between two values, while yet others highlight large differences. Enhancers can be used to further qualify either smallness or largeness. Other expressions designate upper or lower limits on the values observed. Within these there is often a distinction between words used with countable quantities and uncountable measurements.

There is a difference in connotation in the words used to indicate upper and lower limits. In the following examples, using "as many as" implies that 16 is a large enough number, a situation known as permitting, whereas the connotation of "no more than" is that 16 is a small or limiting number, and "a maximum of" is neutral, even though all place the upper limit at 16.

- There can be as many as 16 different options. [permitting]
- There can be a maximum of 16 different options. [neutral]
- There can be no more than 16 different options. [limiting]

One term requires special mention: significant. While often used as an enhancer, for example "significantly larger than expected," this should be avoided in scientific and technical writing, for reasons discussed above (section 22.3.2).

Measurement qualifiers will most commonly be used in the results or analysis:

- As many as 15 occurrences were detected in a single trial, so the results were essentially the same as predicted.
- No more than six occurrences were detected in a single trial, which is appreciably fewer than formerly reported.

In both examples, the first underlined qualifier directs readers' interpretation of the data, while the second indicates a comparison, either with the predicted result or formerly reported results. The adverbs "essentially" and "appreciably" modify the claims of similarity in the first example and difference in the second, making them more defensible or more specific, respectively.

Modals express the likelihood, advisability, or possibility of an action, and are widely used to modify claims.

Modals

Another way to modify your claim or suggestions is the use of modals. Modals can be used to express several different concepts related to the likelihood, advisability, or possibility of an action, as shown in Table 22.4.[13]

It should be apparent that the modals for two of the uses, likelihood and advisability, are nearly identical, apart from "need to." This is logical, as if you tell someone that a choice is necessary, it makes the likelihood of it happening virtually certain; if you tell them that it is advisable, it is probable or likely that it will happen; but if you are saying it is optional, the decision is up to them, that outcome is possible but neither certain nor

Table 22.4 Modals

Use	*Category*	*Modals*
Likelihood	Virtually certain	Must, have to
	Probable/likely	Should, ought to
	Possible	May, might, could, can
Advisability	Necessary	Must, have to, need to
	Advisable	Should, ought to
	Optional	May, might, could, can
Capability, Possibility	Ability	Can
	Permission	May, can

even likely. Capability can be thought of as a subset of the optional choice category, differentiating between whether the choice is a matter of ability or permission.

As with the likelihood qualifiers previously discussed, modals can be used to modify the "truth value" of a claim, as in the following examples.

- The rise in temperature was due to . . .
- The rise in temperature must have been due to . . .
- The rise in temperature ought to have been due to . . .
- The rise in temperature may have been due to . . .

The first example, without a modal, is the basic claim. Stated this way it implies that you have the evidence to back up your claim, and that you are quite certain that this is the only reasonable interpretation of the evidence, so it should be considered a fact. The second implies that you are virtually certain of your conclusion, although some other interpretation might be possible, either because the evidence could possibly be interpreted in another way or because the evidence is limited in some way. The third example has the connotation that you are fairly sure of your conclusion but want to leave open the possibility that there might be other explanations. The last example indicates that you know your interpretation could not be adequately supported by the available data, although it is not excluded.

Next, consider how modals modify the strength of a suggested course of action.

- To avoid this rise in temperature we must . . .
- To avoid this rise in temperature we should . . .
- To avoid this rise in temperature we could . . .

In the first example, the author views the particular action as necessary, because that is the only way the rise in temperature can be controlled. In the second example, the action is stated as advisable, that it will have the desired effect, although it might not be the only way to accomplish it. The final example gives one option but does not claim it is the only or even necessarily the best option to accomplish the purpose. Thus, modals can modify how strongly we advise a certain action.

The final use, capability, can be problematic. Although in common speech it is common to use "can" to indicate permission, in writing it is better to use "can" only for ability, and "may" for permission, to reduce ambiguity. "Can" is often used in a related sense in IPTC papers, to indicate possibility, as in the following example.

- The output can be improved by . . .

One of the difficulties of speaking or writing in a second language is that connotations may differ from one language to another, even though the base meaning may be the same. This is certainly true with modals. The Chinese word usually translated as "should" implies suggestion but not obligation, whereas "should" in English tends to imply obligation.[14] Thus if someone says, translating from the Chinese, "You should wear a sweater in this weather," they may mean it as a friendly suggestion, but someone whose first language is English may hear it as critical of their decision not to wear a sweater. Thus you should be careful when using "should."

Delimiters

> **Delimiters note the limitations of your research and what remains to be done.**

Although your work makes an important contribution to the field, it is not the final word. At the end it is common to note the *limitations of your study and what work remains to be done, either by your lab or others.*

What limits your study? Do you have a limited data set or have you had limited time to analyze the data? Is further work going on in your lab to resolve those questions or do you need the expertise of others outside your lab? Is this the starting point, or an ongoing study, and what is the next step? All of these can be indicated by choosing appropriate delimiters (Supplemental Table 22.14).

Since the research published in any article is only a small part of an ongoing project, which is an even smaller part of ongoing research in your field, a brief discussion of the limitations of your study helps coordinate that effort. By admitting the limitations, you concurrently point to what still needs to be done. By stating what work is ongoing in your lab, you alert others to work that is likely to be published soon. By suggesting that other work needs to, could, or should be done you highlight areas in which others could make a contribution. All of this is usually done in one or a few sentences, as shown in the following examples.

- These conclusions are applicable within the range of our data, but further studies would need to be done to confirm their general validity.
- Ongoing studies in our lab will continue to clarify these remaining issues.
- Similar studies should be conducted with other organisms.
- There is an urgent need for further work in this area.

Delimiters can also be used to indicate that you realize the limitations of your data, that there are possible exceptions to the general rule, or that the results are inconclusive. Examples of this type of delimiters are shown below.

- Based on (limited/preliminary) data, . . .
- The results indicate that (with the exception of/apart from) . . .
- The data (tend to/seem to) imply . . .

Hedges

In addition to softening claims to make them easier to defend, words are sometimes used to *reduce the impact of an utterance on the hearer.* Linguists call this hedging, and attribute it to politeness or cooperation. Like physical hedges, linguistic hedges help ensure good relations with the neighbors, in this case others in your field whose work you are claiming to surpass.

Many of the qualifiers listed above, particularly likelihood qualifiers, can serve not only to reduce the strength of a claim but also to make it more palatable to your readers. Similarly, some delimiters, particularly those claiming limitation of knowledge, serve a double purpose as hedges. For example, when claiming to be first, it is essential to soften your claim, just in case there is a similar paper that you have not noticed, or one comes out after yours is submitted but before it is published, as in the example below.

Hedges soften the impact of a statement on the reader, and overlap with the qualifiers listed above.

- As far as we know, this study is the first of its kind.

One of the most common hedges in academic writing is using words or phrases that indicate uncertainty. Other hedges include words referring to the thought of the researchers or the unproven nature of certain assertions. Advice for others doing similar work is often given in the passive to avoid appearing too directive. Examples are shown in Supplemental Table 22.15, but it is by no means a comprehensive list.

Verb connotations

Verb choice can also change the connotation of a sentence.

One of the easiest ways to change the strength of a claim is simply to change the verb. Just as choice of a certain citation verb can indicate how an author feels about previous research, verb choice can indicate how strongly the author feels about the conclusions of the present paper.[15]

Consider the following examples:[16]

- Our results demonstrate that . . .
- The tests show that . . .
- This finding indicates that . . .
- We conclude that . . .
- The outcome suggests that . . .
- The observation implies that . . .
- It appears that . . .

It should be clear that the first is a much stronger statement, while the last is weaker.

Combining qualifiers

All of the above methods of qualifying your claims can be used separately or in combination. For example, it is possible to begin with a likelihood qualifier and then add some other qualifier to it, as in the following examples.

- It is likely that A caused B. [Likelihood]
- It is likely that A often caused B. [Likelihood + Frequency]
- It is likely that A caused a substantial change in B. [Likelihood + Similarity]
- It is likely that A may have caused B. [Likelihood + Modal]

> The possibility of combinations of qualifiers is almost endless, so do not expect to master them all immediately.

Examples of some of the more common hedges found in technical writing are shown below. All of these could be added to the beginning of the previous examples to further soften the claim. However, as with achievement markers, using too many in the same place can ruin an article.

- It (seems/appears) that . . .
- This (suggests/points to) . . .
- It can be (assumed/inferred) that . . .

Exercise 22.3 Using qualifiers to make your claims more supportable

(1) What qualifiers are used in your exemplar articles?

Hint: Consider both the types of qualifiers used and where they are used in the article.

(2) Does your usage match? Are there places you should add qualifiers to soften your claim or clarify your data or opinion of other research?

Evaluate your understanding

(1) What is the purpose of adding connectors to your writing?
(2) How are contribution markers, achievement markers, and application markers similar? How are they different?
(3) What devices can provide connection between sentences?
(4) Why are there so many logical connectors, when the same basic ideas could be conveyed more simply by coordinating conjunctions?
(5) What is one of the most important uses of connotation in science and engineering research articles?
(6) What commonly used terms should be avoided in research articles, unless used in their technical statistical sense?
(7) What types of qualifiers can be used in summarizing data as well as clarifying claims?
(8) What is the purpose of using modals, delimiters, and hedges?

Homework

(1) Rewrite your draft again, concentrating on clarifying your argument using the markers, connectors, and qualifiers described in this chapter.

Hint: Look at the Supplemental online material for further details.

Hint: Again, at this stage you may want to show your draft to your instructor or advisor and get comments on it before you do the final revision.

Teaching tip: This chapter is long, but has been kept together because both connection and connotation are related to clarification. Connection and connotation can be taught as separate classes, or some subsections covered with either the previous or following chapter, both of which are much shorter.

Supplemental material:

Material particularly useful to ELF students

Class exercise 22.1—logical connectors (with answer key)

Notes

1 Conciseness is less important in some social sciences and humanities.
2 Glasman-Deal (2010) heightened my awareness of the importance of these connections and her word lists were invaluable in developing the supplemental tables for markers, although my analysis is far different.
3 The verbs and adjectives listed in the online supplemental material could be thought of as potential contributions or achievements, respectively.
4 All examples of continuity connectors are from Rau (2012), emphasis added.
5 Modified for example.
6 Steps in an argument can be ordered "first, second, . . ." or "firstly, secondly, . . ." but not "at first, . . ." which implies a time sequence rather than a logical sequence. "At" is not used with ordinal numbers above first ("at second") in American or British English, although this is common in Vietnamese English.
7 Grammarians differ on whether "but" can be used in sentence-initial position; some say that since it is a conjunction, by definition it *should not be* used to begin a sentence, while others point out that it *is* used that way in many written documents and can be treated as a connector as well as a conjunction.
8 Properly the word "data" is plural (singular: datum) and takes a plural verb. Some journals accept data as collective, with a singular verb.
9 Change in thought in the field would be common in the Introduction, while sequence of events in the research may be found in the Process and Testing (Methods and Results).
10 I don't know if the law student was bored, but the author was a law student. The examples are from www.tylervigen.com/spurious-correlations.
11 Likelihood and probability are distinct mathematical concepts, but the distinction is hard to understand for those outside the field, and the terms are used interchangeably in most disciplines except mathematics and philosophy. The other two have no formal mathematical expression.
12 Biber, Connor, and Upton (2007: 69–72) show many ways of expressing the authors' stance, many of which relate to how confident they are about a statement.
13 This analysis is simplified from Glasman-Deal (2010: 158–167).
14 應該 vs. 必須 (simplified characters: 应该 vs. 必须; pinyin: ying1gai1 vs. bi4xü1).

15 There is substantial overlap between these verbs and those used to express an evaluation of past research, listed in Table 7.3. Both verbs and adjectives can be used in different ways to indicate what linguists call stance, including how certain we are about our argument. See Biber, Connor, and Upton (2007: 69–72).
16 Cargill and O'Connor (2013: 62) give more ideas on using sentence patterns employing a "that" construction to modify the strength of a claim.

References

BIBER, D., CONNOR, U. & UPTON, T. A. 2007. *Discourse on the move: Using corpus analysis to describe discourse structure*, New York, John Benjamins Publishing.

CARGILL, M. & O'CONNOR, P. 2013. *Writing scientific research articles: Strategy and steps*, New York, John Wiley & Sons.

FRANKLIN, B. 2013. *Poor Richard's Almanack and other writings*, Chelmsford, MA, Courier Corporation.

GLASMAN-DEAL, H. 2010. *Science research writing for non-native speakers of English*, Singapore, World Scientific.

HU, L., KU, W.-S., BAKIRAS, S. & SHAHABI, C. 2013. Spatial query integrity with Voronoi neighbors. *IEEE Transactions on Knowledge and Data Engineering*, 25, 863–876.

KLEINHEINZ, G. T., MCDERMOTT, C. M., HUGHES, S. & BROWN, A. 2009. Effects of rainfall on *E. coli* concentrations at Door County, Wisconsin beaches. *International Journal of Microbiology*, https://doi.org/10.1155/2009/876050.

RAU, G. 2012. *Mapping the origins debate: Six models of the beginning of everything*, Downers Grove, IL, InterVarsity Press.

ROTH, P. 2011. *The ghost writer*, New York, Random House.

SWALES, J. M. 1990. *Genre analysis: English in academic and research settings*, Cambridge, Cambridge University Press.

Chapter 23

Revision and finalization

23.1 Time's up

There is no fixed endpoint for rewriting. If you wait a little longer, work a little more, you will always be able to improve your document. Perhaps it is counterintuitive, but the better you get at writing, the truer that will be. So how do you know when to quit? Sometimes you reach a point where you are making smaller and smaller changes, and just feel enough is enough. In many cases, it is just that time's up. The paper is due, the deadline is here, or you have to begin work on something else. As you read it one last time you may feel it still does not express exactly what you want to say. Nevertheless, the time has come for the final revision and proofreading.

This chapter concludes the topic of rewriting using the 7Cs of Change, covering the stages highlighted in Table 23.1. For consistency and correctness you will be making small changes, mostly at the level of words or formatting, but collaboration may lead to changes at any level of the document. Before or during this step you will also need to finalize the graphics and add the Title, Abstract, and whatever other finishing touches are required (Chapters 24–26).

23.2 Consistency

We will consider two areas where consistency is important: thought and words. I will also talk in this section about removing ambiguity, by making sure all your referents are clear and correct. The best way to see whether the document is consistent is to read quickly through the whole thing from beginning to end, focusing on the main thoughts and

Table 23.1 The 7Cs of Change, highlighting those covered in this chapter

Type of change	*7Cs*	*Purpose*	*Level*	*Scope*
Rewriting	1 Coherence	Organization	Paragraph	Section
	2 Conciseness		Sentence	
	3 Connection	Clarification	Sentence	
	4 Connotation		Words	
Revising	5 **Consistency**	**Finalization**	Words	Whole paper
	6 **Correctness**		Format	
	7 **Collaboration**		All	

keywords rather than details. I conclude the section with a warning against employing the squid technique.

23.2.1 Consistency of thought

> Authors frequently restate the research goal in the final rewrite, but must check for consistency.

It is very common to finish your research and find that you have not achieved the research aim or answered the research questions you wrote at the beginning. Instead of starting over, it is simpler to reformulate or rephrase the research goal. Remember, however, that your argument is entirely based on that goal. Thus you also need to change any references to that goal in the Process or Testing sections and the contribution or answers to the research questions in the Conclusion.

I vividly remember my first year in college writing a paper on memory. At that time it was thought that some class of chemical was responsible for memory. I researched the most recent findings, came up with my conclusion, put my note cards in order, and started typing. When I reached the final note card, very close to the deadline, I realized that final bit of data disproved my hypothesis. To incorporate it I would have to rewrite and retype the entire paper. So I did the only reasonable thing—I threw the card away, concluded the paper without including that reference, and turned it in. When I got it back the professor had commented, "I am surprised that a freshman has sufficient background to write such a paper." I laughed, relished my A, and went on to the next course.

Unfortunately, following my example is not an option with your research. If you find at the very end that something is wrong, you will have to go back and change the whole paper. Fortunately, that is much easier with personal computers than it was in the days of typewriters. Unfortunately, the ease of making changes also means that editors tend to require more changes than they did before.

The longer it has taken to complete a work, the more likely it is that your thoughts have changed. You may have changed your terminology or substituted a new design element, so it is essential to read through the whole paper carefully, checking for consistency of thought. While writing this book I went through several different versions of the originally 6Cs and later 7Cs of change. In the final revision I had to make sure they were listed properly throughout the book. You will need to do the same.

23.2.2 Consistency of words

> Technical terms should be used consistently, but other words can be varied to maintain reader attention.

Problems with consistency often occur at the level of words. By using the same word consistently, the reader knows you are referring to the same concept. At the same time, keeping the reader's attention requires variation to avoid monotony.

When deciding whether to repeat the same word or use a synonym, the main question to ask yourself is, *is this a keyword or specific technical term, or is it a general concept that could be expressed in different words?* Technical terms must be used consistently, as no synonym expresses the same thought. Apart from those, however, it is good to vary your word choice when you can.

The first step, then, is to identify the keywords. Think about the words I have used in this book. I have been careful to distinguish division from section and have not used either term for any other portion of an article. Nor did I use any other word for component, or use that word with any other meaning. Such words must have a consistent meaning and be used without variation for clarity. As you consider your writing, the keywords may be physical apparatus or concepts. A few may be verbs—processes with a specific technical name. These must be referred to the same way every time.

For other words, consider what synonyms could be used, especially if the same thought appears several times in one paragraph. This is particularly true of the first two divisions, where you report what others have done and what you did, both of which can be very repetitive. Remember, however, that your primary goal is clarity. Do not use any word if you are not 100% sure of its meaning. If you use the wrong word, or one that does not make sense in the context, the reviewer will undoubtedly see it for what it is: someone who made use of a thesaurus in a failed attempt to sound erudite.

Search your document for both old and new forms of any terms you have changed to make sure they are used consistently. Similarly, if you think one word occurs too frequently, search for all occurrences before deciding which would be the best to change, rather than just changing some randomly.

Exercise 23.1 Consistency of words

(1) Check your writing to make sure key terms are used consistently. Apart from those words, if the same word appears twice in a paragraph, consider whether it would be possible to substitute a synonym.

Hint: If you notice any problems, check for occurrences elsewhere in the paper.

23.2.3 Eliminating ambiguity

Many different types of words or expressions can be ambiguous, usually because the referent is not clear. It is important to ensure that your reader understands every sentence the way you intended it. While novelists may purposely use ambiguity early in a story to set up a surprise ending, that technique is not appreciated in the scientific literature. Gifted writers might be able to pull it off when writing popular science, but it is extremely rare in academic journals. Common sources of ambiguity include abbreviations, pronouns, antecedents, and modifiers.

> Abbreviations, pronouns, antecedents, and modifiers all need to be checked to eliminate ambiguity.

The general rule for abbreviations is that the full name must be stated along with the abbreviation the first time it is used. While this works well for research papers for class, it is not always sufficient for research articles. On the one hand, certain abbreviations are so well known in a field that they do not need explanation, like DNA in biology or LED in electrical engineering. On the other, since readers may not read the whole paper it can be helpful as you begin a new section to reintroduce uncommon or new abbreviations. There is no hard and fast rule for this, but you need to be aware of what your audience knows and does not know. Expectations for abbreviations in an Abstract differ, so be sure to check the requirements.

Pronouns refer to something already mentioned, frequently but not always in the previous clause. In practice, relative pronouns may also be used to refer to the entire previous sentence, as in the following examples, although prescriptive linguists may feel this is an incorrect use:

- This is mainly due to the distance between . . .
- This was not borne out by extensive . . . field measurements.

Misplaced modifiers are the bane of a writer's existence and can particularly be a problem with very concise writing, including headlines, section titles, or titles of graphics. Consider this ambiguous headline from BBC News:

- Liquid living worms survive space (30 November 2011)

As a biologist, I know there are no liquid worms, and it is hard to imagine how solid worms could be liquefied and survive anywhere. In this title it initially appeared that both "liquid" and "living" were individually modifying "worms." A hyphenated form, "liquid-living," would have resolved the ambiguity, making it clear the worms lived in liquid.

Or consider this headline:

- EU waters a cemetery, Malta warns (12 October 2013)

This made me wonder why the EU would want to water a cemetery, and why Malta would care if they did. Later it became clear that Malta, an island in the Mediterranean, was warning that the waters around the EU *are* a cemetery, devoid of life. Here the problem is that "waters" could be interpreted as either a noun or a verb.

One more, this time reported by the editors of the Columbia Journalism review:[1]

- Cause of AIDS found—scientists

Colons and dashes can serve many functions. Make sure your intention is clear.

The best way to check whether your writing is ambiguous is to ask someone not familiar with your research to read and comment on it. Since you know the background and have thought about the subject extensively, you will tend to fill in information that is not actually written in the paper. Since you wrote it, you will know whether you intended a word to serve as a noun, adjective, or verb. As much as possible, put yourself in the position of someone reading the paper for the first time. What information would you need to know, what referents are not clear, or what words could be interpreted as a different part of speech, leading to confusion?

23.2.4 Squid technique

A different type of ambiguity is often referred to as the squid technique, so named by Doug Saville, who said,

> occasionally, I recognize what I call the squid technique: the author is doubtful about his facts or his reasoning and retreats behind a protective cloud of ink.[2]

> The squid technique involves intentional use of ambiguity to try to hide the fact that you are not clear about some aspect of your research. Don't do it.

Just like a squid trying to confuse potential adversaries, authors can be purposely ambiguous in an attempt to hide the weakness of their argument. It may work for a squid, but rarely for writers. If you can clarify your writing with a little additional work, do it. If it requires further research that has not yet been done, say so.

For example, do not say, "It has long been known that . . ." if what you really mean is "I was too lazy to find the reference." Look it up!

Do not say, "Two thirds of the samples showed the same pattern," if you only had three samples. Do the work required to get quantitative results you can truly stand behind or use a qualitative approach to the analysis.

Do not say, "obviously" or "clearly" if what you really mean is, "I really don't want to list all the steps in my logic; in fact, I am not sure of all the steps myself." Make your argument clear enough that your readers find it obvious themselves, without you saying so.

Finally, do not say, "Much more work will be required to provide definitive answers to these questions" if you really mean, "I don't have a clue how to explain the data." Emphasize what you *have* learned through the present study that others can build on, not the questions that are not yet clear. If there is nothing, you really have nothing to publish.

23.3 Correctness

The final stage of revision involves one final check that everything is correct. What you need to focus on at this stage depends on your background. Ironically, those who learned English as a first language may make more mistakes with certain frequently misused words than those using English as a lingua franca, who may have read and perhaps even memorized such word lists. On the other hand, those using English as a second language will have more difficulty with little words like prepositions and articles. Both need to be aware of differences between British and American English and take the time to proofread carefully for misspellings and formatting errors.

When checking for consistency, I suggested that you read the whole article quickly. To check for correctness I suggest the opposite: that you read it again slowly. Some people prefer to read through the article backwards one sentence at a time to make sure they are not distracted by the content. While experienced editors may be able to check all the different categories in this section at the same time, you may find it helpful to focus on one or a few at a time.

23.3.1 Frequently misused words

Many websites list frequently misused words.[3] In this section I will mention only a few words that are more likely to be found in scientific, engineering, or technical writing.

> Some word pairs are frequently used incorrectly by those using English as a lingua franca, others by those whose first language is English.

Some words that are frequently misused are linguistically related, with similar-sounding words for the noun and verb. In many cases the two can be interchanged in your writing by changing the structure of the sentence, as shown in the examples below.

Advice (n)/advise (v): When your advisor advises you, he or she gives you advice.
When I want your advice, I will ask for it.
When I want you to advise me, I will ask you to do so.

Device (n)/devise (v): A device is something that someone has devised.
Our solution to the problem is a workable device.
We devised a workable solution to the problem.

Effect (n)/affect (v): A variable strongly affects the results if it has a large measurable effect.
Variable A had a strong effect on the results.
Variable A strongly affected the results.

Other words have similar meanings that are easily confused, whether they are linguistically related and sound similar or not. These cannot be substituted for one another, as they are used to refer to different categories of objects.

Continuous/continual: continuous means without interruption and has a broader meaning, referring to numbers or space as well as time, while continual is used only with respect to time and often implies an interruption in the continuity, so the event is regular rather than continuous.
Variable A is continuous, while variable B is categorical.
We had to deal with continual breakdowns in the apparatus.

Each/every: each refers to separate members of a group of two or more, while every refers to all the members of a group of three or more.

Each time we did the test we did it the same way.
Every time we did the test we got the same results.

Farther/further: farther is used only with respect to physical distance, whereas further is used when referring to abstract or metaphorical distance.

We went farther from the shore than we did in previous studies.
We have gone further in this work than previous authors.

Fewer/less: fewer is used with countable objects, while less is used with uncountable objects or numbers referring to measurement or time.

We collected fewer samples than we did the first time.

We collected less water than we did the first time.

We collected less than 2 ml of water this time.

Imply/infer: something implied is not specifically stated, whereas something is inferred if it is deduced from the evidence.

We implied the previous results were wrong, but did not say so explicitly.

We inferred the previous results were wrong, based on our new evidence.

Yet another category of words that are commonly misused are homonyms: words that sound almost the same although they have very different meanings. In some cases, only one is commonly found in the scientific literature.

Complement/compliment: a complement makes something complete, either physically or metaphorically, while a compliment is praise given to a person.

The IPTC model complements the IMRD model, but does not replace it.

I hope to get at least a few compliments on my work.

Discrete/discreet: items that are discrete are separate, but things that are discreet are handled carefully and confidentially.

There are 100 discrete points on the graph, each representing a person.

We have to be discreet not to reveal the identities of the individuals.

Elicit/illicit: to elicit is to draw something out of a person, whether information or a reaction, but illicit is that which is not acceptable under the law or conventional moral standards.

We tried to elicit responses from the subjects, but had limited success.

They were unwilling to talk about their illicit activities.

Idea/ideal: an idea indicates a suggestion, purpose, or plan of action, whereas ideal means the best.

I have an idea how to solve this problem.

It may not be the ideal solution, but it should work.

Its/it's: its is a possessive pronoun, meaning belonging to it, whereas it's is a contraction of it is.

It's considered improper in many journals to use contractions like "it's."

If you are not sure about its usage, see if substituting "it is" makes sense.

Lose/loose: lose, used only as a verb, is the opposite of find, whereas loose, the opposite of tighten or tight, can be used as a verb, adjective, or occasionally noun.

If you set a helium balloon loose, you will lose it.

If you lose it, it will be lost forever.

Principle/principal: a principle is the fundamental foundation of a set of beliefs or a theory, while principal means the main or most important item in a group.

One principle of science is that whoever publishes first gets credit for a discovery. That is why time of publication is of principal interest to researchers.

Another frequently misused class of words are those that have irregular plural forms, usually because the plural form of the original Latin or Greek is retained. In addition to the words listed below, many terms in biology have irregular plurals.[4]

Datum (s)/data (pl): a single datum is frequently called a data point, while the word data is a plural form.[5]

The data are presented in Figure 1. The first data point represents the initial conditions.

Analysis (s)/analyses (pl).

Previous analyses have underestimated gene number. The present analysis corrects this problem.

Criterion (s)/criteria (pl).

The first criterion for choice of a site is accessibility, but other criteria may prove more important in the long run.

Phenomenon (s)/phenomena (pl).

By definition, science seeks to explain natural phenomena, but data give evidence of a phenomenon, so in practice science seeks to explain data.

One final word that is often used improperly is respectively. Respectively is used to indicate that two lists are given in the same order, and is always set off from the rest of the sentence with a comma or commas. It cannot be used if there is only one list, nor can it be used if the number of objects in each list is different. It is also never used to indicate that things happen sequentially.[6]

Correct uses:

The two subjects were male and female, aged 67 and 63, respectively.

The two subjects were, respectively, male and female, aged 67 and 63.

Incorrect uses:

The two subjects were aged 67 and 63, respectively. [No second list.]

The three subjects were male and female, aged 67, 63, and 62, respectively. [Unequal numbers so it is unclear which were male, which were female.]

The two subjects were a male aged 67 and a female aged 63, respectively. [Two lists are already merged, so there is no need to add "respectively."]

The two subjects will be described respectively. [One after the other = sequentially, not respectively.]

23.3.2 Prepositions

One of the greatest challenges for anyone learning a second language is to master the little words. Whether the prepositions and articles of English or the measure words of Chinese, rules can be stated but there always seem to be exceptions. One underlying problem is that no two languages have an exact correspondence of how they categorize thoughts.[7] Another problem is that language is always changing, so British English may differ from American English and there is even variation between regions, or based on class, age, or gender. Moreover, the form used in a given area in common speech may not be considered acceptable in academic writing, which is more formal and conservative than speech or even popular writing.

With prepositions, the first issue is that the meaning of some verbs is incomplete or different without the preposition. For example, in English we "get *in*" the car. Note that *get* has a very different meaning without the preposition *in* and that no other preposition is used in this context. In Chinese, however, to express the same meaning one would say "get *on*" the car, undoubtedly because the Chinese word for "car" can mean any wheeled vehicle, and people got *on* wagons long before there were cars.[8] While such connections between verbs and prepositions are naturally internalized by those who learn the language as a child, those who learn a second language later in life need to memorize the patterns, since there is no way to transfer expectations from one language to another.

> Prepositions are a major problem since languages change over time, leading to variation and many different meanings assigned to a single word.

A second problem is that often the same preposition is used to express various meanings, something few native speakers of a language will be aware of.[9] For example, consider the following sentence:

> *In* the morning I found the information *in* an article *in* a folder while getting *in* the car, so I will give it to you *in* an hour.

In this sentence, *in* is used in five ways, first referring to a time of the day, then a metaphorical use, next physical space, then meaning into, and finally meaning within a certain length of time from now.

Since there are so many online resources on prepositions, many of them specific to problems faced by speakers of a particular language, I will not dwell on them here. A few expressions that seem to be common problems in scientific writing are worth mentioning, however, particularly those dealing with comparisons.

Starting with the basic comparison, should you *compare with* or *compare to*? In most cases, either is acceptable, as long as the comparison is between real objects. For example, you could compare your data *to* previous data, or compare your design *with* existing designs. If you are comparing in some way that is metaphorical or based on resemblance or analogy, you must use *to*, for example when comparing *to* an ideal. Since the ideal does not actually exist, you cannot compare *with* it, which often implies the ability to lay the two side by side for comparison.[10]

The next question is what preposition to attach to comparatives like *same* or *similar*. Some Chinese seem to think that because you compare *with*, the same preposition should be appropriate for the resulting judgment about sameness, resulting in *same with* or *similar with*. Logically this might make sense, but unfortunately neither of these are considered correct in either American or British English, which prefer *same as* and *similar to*.[11]

Even more difficult is what preposition to use with *different*. In both British and American English different *from* is the most common, while different *with* is considered improper, but in Britain different *to* is also acceptable, while America also uses different *than*. Reasons for this are vague, but different *from* will work in most situations.[12]

Another common question is whether to use try *to* or try *and* in sentences like, "We will try *to/and* demonstrate that our product is superior to any currently on the market." Some authors, particularly in America, unambiguously say that try *and* is grammatically incorrect. Others label it an idiom or fixed expression and say it is fine. Yet others suggest that it is fine for colloquial speech but should be avoided in academic writing.[13] Since try *to* is universally accepted but some people frown on try *and*, stick to the former.

For those whose first language is not English, a good rule of thumb for using prepositions is if referring to graphics, use "in," but if referring to how you collected your data or did your analysis, use "by." This will be a good start in many cases. If you want to see how other authors use prepositions, you can search your exemplars for a certain word and see which prepositions are used after it.[14]

23.3.3 Articles

Those who have grown up using articles cannot appreciate the difficulty they pose for those learning English as a foreign language. In general, definite article *the* or indefinite article *a(n)* are used with singular nouns, while with plural nouns the choice is between *the* and no article [Ø], since *a(n)* implies singularity. Changing the article will change the meaning of the sentence, as in the following.

This may reveal *the* connection between the two. (a specific connection)

This may reveal *a* connection between the two. (some connection)

Attach the wires to *the* two adjacent sites. (specific adjacent sites)

Attach the wires to [Ø] two adjacent sites. (any two adjacent sites)

The flowchart in Figure 23.1 attempts to summarize how to determine which article should be used. Proper nouns, the name of a person or location, are obviously specific, so it seems that they should always take a definite article. Nevertheless, names of people, states, cities, mountains, and lakes normally take no article. Other bodies of water like oceans and rivers, along with buildings, on the other hand, do. Don't ask me why. I don't make the rules. For example:

Queen Elizabeth II

Minnesota

London

Mount Everest

Lake Chargoggagoggmanchauggagoggchaubunagungamaugg[15]

The Atlantic Ocean

The Nile River

The Louvre Museum

For common nouns, the first question is whether it refers to some specific item known to the reader. Things that have been previously mentioned or can be inferred based on what has been mentioned are known, therefore the definite article *the* is used.

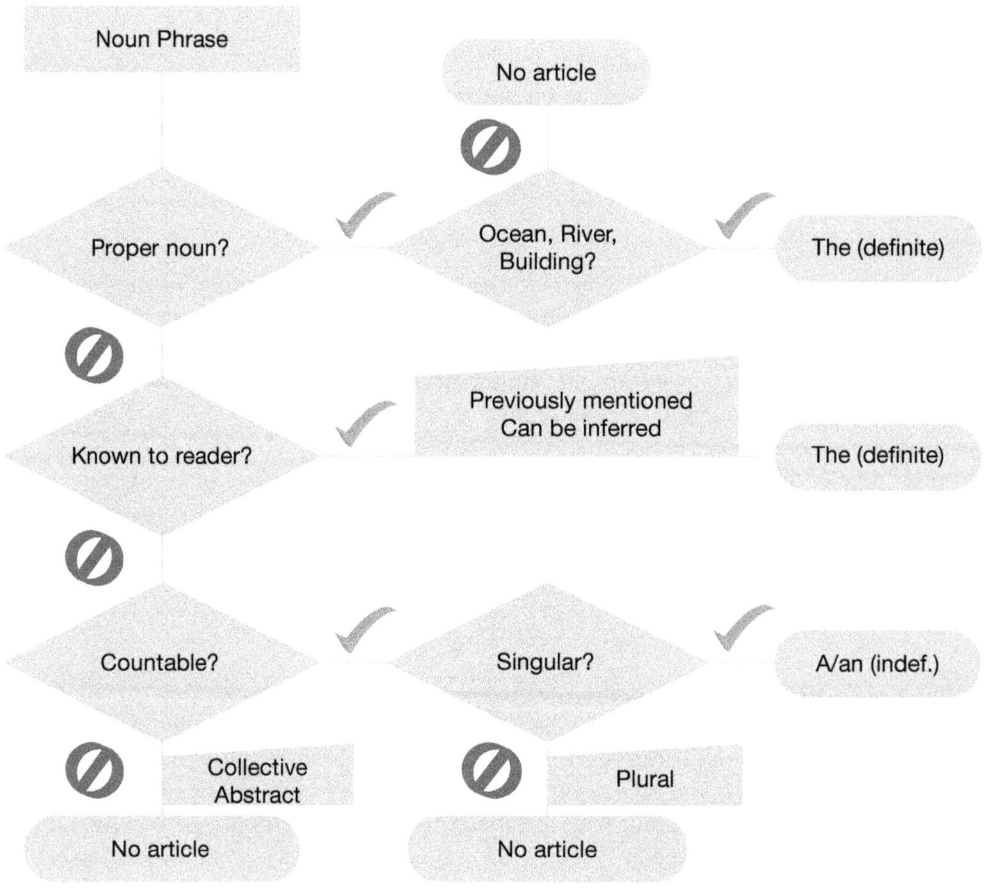

Figure 23.1 Flowchart for deciding what article to use.

To illustrate the importance of what the reader knows, consider the following.

> I had *a* bowl of beef noodles for lunch.
>
> *The* noodles were very good, but *the* beef was tough.
>
> *The* broth was also very salty.
>
> So I did not drink *the* broth, but bought *a* glass of tea.

In the first sentence *a* is used because this is the first mention of the subject. In the second sentence, *the* is appropriate because both the beef and the noodles have already been mentioned. In the third sentence *the* is also used, even though it is the first mention of the broth, because I can assume that the hearer will make the connection between the beef noodles and the broth they are served in. In the last sentence, the tea is again a new topic, so I use *a*.

The is also used if the hearer can easily infer which of several possible referents is meant. On the other hand, *a* is used if the specific referent does not matter or is unknown to the hearer. For example, I could say to someone in my department,

> I ate lunch with *the* department head.
>
> He had just come from *a* meeting in *the* administration building.

In this case, the hearer will infer I am referring to our department head, not some other. Similarly, although there are many administration buildings, there is only one on our campus, so the referent is clear. On the other hand, there are many different meetings and I cannot assume the hearer knows which one I am talking about, so *a* is appropriate.

If the noun phrase is not a proper noun and not known to the reader, the next question is whether it is countable. Collective and abstract nouns never take an article. Unfortunately, determining whether a noun is countable or not is not always easy. Most concrete nouns that refer to physical objects are clearly countable, such as organism(s) or circuit(s). Yet some nouns that refer to objects do not have a plural form, such as equipment, and are considered collective.[16]

An even greater problem is that many nouns can be considered either countable or uncountable depending on their use in the sentence. For example, absence is an abstract concept that is uncountable, but absences from class are countable. Paper can be used as a collective term for a type of material (sheets of paper),

Frequently whether a noun is considered countable or uncountable depends on its context. Other rules for article use likewise have exceptions.

or as a countable term (two conference papers). One book on scientific writing lists work, freedom, and intelligence as uncountable nouns, and often that is true, but we can also talk about how many works an author has produced, the freedoms of speech and religion, or multiple intelligences. To make it even more confusing, American and British English may differ. Any rule or list I could give would have exceptions, so you must consider whether the word could be considered countable as you are using it. If English is not your first language, just accept that it will take many years to master this.

Singular countable nouns that are unknown to the reader take the indefinite article *a* or some other "determiner" like *my, this, some,* or a number. For plurals, the article *a*, being singular, cannot be used, but other words can be. For example,

> We submitted *a/the/my/this/one* paper to the journal *Nature*.
>
> We submitted *the/our/these/two* papers to the journal *Nature*.

One special case important in technical writing is when the adjectives *few* or *little* are used to modify the noun. Since these are used with plural or uncountable nouns, respectively, we would expect no article. Nonetheless, *a* can be inserted to modify the meaning of the sentence. Without an article, the connotation of the sentence is the rarity, whereas with the article the emphasis is on the presence of the noun.

> We found few exceptions to this rule.
>
> We found a few exceptions to this rule.
>
> This causes little trouble for native speakers.
>
> This causes a little trouble for native speakers.

One final rule refers to when to use *an* instead of *a*. The rule is sometimes stated in terms of the following letter but is better expressed in terms of the following sound: Use *a* if the following sound is a consonant but use *an* if the following sound is a vowel. Thus words like university and unique, which are considered to begin with a consonantal y sound, take *a*. This is also true with abbreviations. Again, the only odd case among the vowels is U, while consonants that take *an* include F, H, L, M, N, R, S, and X.

> I teach at *a* university.
>
> We used *a* UV light.
>
> We performed *an* MRI scan.

Exercise 23.2 Articles

(1) In your exemplar article, circle all the *a, the*, or places in front of a noun that could have an article but do not [Ø], and try to determine which rule is being followed.

Hint: Consider the background of the authors when deciding which exemplar to examine for this exercise.

23.3.4 British vs. American usage

Linguists speaking of English as a lingua franca talk about multiple Englishes and try to describe the distinctive characteristics of each, not only in countries where it is spoken as a first language, like Singaporean English, but also places where it is spoken as a second language, like Chinese English. When teaching oral communication, the emphasis in many cases is on communicative competence rather than seeking native fluency and accuracy. Nevertheless, when it comes to written academic communication, British and American standards still dominate. In the past (not the distant past—I still remember it) authors were expected to use one or the other depending on where a journal was published. Now most journals will accept either, as long as you are consistent throughout the paper, but you will need to check the Instructions to Authors to be sure.

> **British and American English differ in word choice, spelling, punctuation, and numbers, but most journals now allow the author to choose either system.**

A surprising number of words differ in British and American usage, most being nouns referring to common objects in daily life. Examples include motor vehicles (lorry/truck), parts of cars (bonnet/hood, boot/trunk), words related to trash (dustbin/wastebasket), food, clothes, holidays, and others. There are a few names of chemicals, drugs, or supplies that differ (aluminium/aluminum, cotton wool/absorbent cotton), but fortunately most words used in scientific or technical writing are identical.[17]

There are also a number of spelling differences between British and American English.[18] Just as the presence of *c* or *s* distinguishes noun from verb in advice (n)/advise (v) and device (n)/devise (v), Britain distinguishes practice (n)/practise (v) and licence (n)/license (v), while America has merged the two parts of speech in different directions to practice and license. Another spelling difference important in science and engineering is that in Britain you analyse while in America you analyze. There are also many words where the –re in British English has been replaced by –er in American English, including litre, metre, and centre. In Britain many words contain double vowels which in America have been replaced by a single vowel, such as colour, behaviour, amoeba, foetus, and paediatric.

Moreover, some words are considered to be collective in American English, but not in British English. For example, in America math is considered a singular or collective noun, while Britain speaks of maths (which makes far more sense, since geometry is as distinct from calculus as biology is from physics). Similarly, in America research is exclusively a collective noun and used with a singular verb, "research has shown," whereas in Britain it is acceptable to say, "researches have shown." Many speakers of English as a lingua franca also speak of researches, so it will be interesting to see whether this becomes acceptable in American journals in the future.

There are also differences when it comes to numbers, as shown in Table 23.2. Due to differences in the meaning of one billion and the date format, the meaning should always be stated unambiguously in international English to avoid confusion. It is also worth

Table 23.2 Differences in number use in British and American English

	British	*American*
One billion	1,000,000,000,000 (1 million million, 10^{12})	1,000,000,000 (1 thousand million, 10^9)
12/1/16	12 January 2016 (day/month/year)	December 1, 2016 (month/day/year)

noting that some countries, including most European countries and some others, use the comma as a decimal mark and either a space or period as a thousands separator, for example, writing ten thousand dollars as $10.000,00.

One final difference between British and American English is the placement of quotation marks relative to punctuation. In British English, punctuation is placed within quotation marks only in cases such as a sentence, where the punctuation occurs in the quoted material, whereas in American English punctuation is always placed within the quote.

> In British English, the rule is "punctuation after".
>
> In American English, the rule is "punctuation within."

23.3.5 Spelling, punctuation, and formatting

Finally we come to the stage of revision commonly called proofreading. The term proofreading originated in the days before computers, when a typed manuscript was physically mailed to an editor, and the journal then had to typeset it on the machine that would be used to print the actual pages. Before final publication a "proof" copy was sent back to the author to make sure that there were no errors in the typesetting process. This still happens with books but is rare for journal articles. Nevertheless, the same process has to be done on your manuscript before you send it to the editor. English editors can do much of this for you, if you are willing and able to pay them, but remember that your name is on the paper, not theirs. The final responsibility is yours.

> Errors due to autocorrect are harder to detect than misspellings, as they are not marked, so it is important to proofread carefully.

Make sure you check your document not just for misspelled words, but also incorrect ones. I had one student who was *quiet* confident of his analysis. Another purchased *regents* for his research. Yet another throughout his whole paper talked about *addictive* manufacturing rather than additive. In *resent* years I have also had papers on *witch* frequency resulted in a *lager* capacity. Such unintentional humor is best avoided.

In section 18.6 I suggested things worth doing right the first time to reduce changes during proofreading. At this point it is still necessary to check all those items, and more.

Similarly, templates help but will not correct all errors. Figure 23.2 lists items you will need to check. Changing the physical appearance of a document by changing the margins, font, or font size can sometimes reveal problems you would otherwise miss. Using the "Show all" command in Word to reveal hidden characters such as paragraph marks, breaks, and spaces is also very helpful.[19]

In addition to whatever is listed in the Instructions to Authors, there are frequently unstated expectations you can learn from your exemplar articles. Hopefully you have already confirmed that the length is appropriate, but at this point you will need to double-check the overall structure, numbering, headings, number of references, and other broader features.

23.3.6 Variation in an ELF research world

It is worth noting that as more and more scientists from non-English-speaking countries write using English as a lingua franca (ELF), expectations are changing. While some journals still expect all authors to conform to the standards of American or British English, other journals are showing greater acceptance of other variants.

Some journals, especially in engineering, are accepting papers with what previously would have been called non-standard English. If you skim some articles quickly you will find that it is fairly easy to understand the thread of the argument, although individual sentences are not formally grammatically correct. This is not always the case, however, and many journals still require that a paper be edited by a "native speaker," preferably one familiar with your field.

23.4 Collaboration

In the collaboration process you ask others to comment on your writing. Although this is placed at the end of the list of 7Cs, in practice the timing depends on what you are writing and how much experience you have had with academic writing. When you are writing your thesis, many advisors suggest that you show it to them chapter by chapter. For your first article, which probably will be written with your advisor as coauthor,

Checklist for final proofreading

Spelling
Double spaces
Foreign punctuation or other marks
Font and font size
Margins, indents
Format of paragraphs—extra line or space before/after, indented or not
Format of headings, titles, captions, footnotes
Format of reference and citations

Figure 23.2 Checklist for final proofreading.

> Collaboration may occur at any stage, not only at the end. For beginning writers it is good to get feedback from your advisor or coauthor early.

you may also collaborate as you write. For articles beyond the first, you may not show it to your coauthor until you think it is finished. No matter when that collaboration takes place, plan on revising your work again afterwards, unless your coauthor does the rewriting for you. In other words, just when you thought you were finished, you have to start all over again. Nevertheless, if you have followed all the suggested stages of rewriting and revision, you should have less work to do after collaboration than you would otherwise.

Remember that revision means looking again, which is what happens when a second set of eyes looks over your document, and those eyes will see things that you missed, whether the eyes are your advisor, another coauthor, or a reviewer. If as you advance in your career you do not have a second author for your paper, but still want comments before you send the paper to a journal, the Internet provides opportunities not available to past generations. Several websites now allow you to exchange peer review with others in your field.[20]

The most important point I can make about collaboration is to learn from it. Look carefully at the changes your collaborators or reviewers suggest and learn from them. Ask for clarification if you do not understand why they made certain changes. Pay attention to common problems identified and try to figure out which of the 7Cs they relate to, so you know where you need to focus your effort next time.

Evaluate your understanding

(1) Why do many writers restate the research goal when the paper is nearly complete?
(2) What techniques can be used to check for consistency?
(3) Which of the many points of correctness mentioned in this chapter is most difficult for you?
(4) As a first-time author, why is it important to seek advice from your collaborators early in the writing process?

Homework

(1) Do a third rewrite of your draft, concentrating on consistency and correctness of the words.

Hint: If you are working on this for a class, this will be the final revision. When you write a journal article, you will have to revise it at least one more time after collaboration before it is ready to submit.

Teaching tip: This chapter may or may not be the focus of your class, depending on how concerned you are with structure and content vs. English usage.

Supplemental material:

Class exercise 23.1: Eliminating ambiguity

Class exercise 23.2: Articles (with answer key)

Notes

1 www.cjr.org/the_lower_case/headlines_editors_probably_wish_they_could_take_back_21.php (accessed 2 September 2016).
2 Quoted in Day (1998).
3 One particularly good site for misused words is www.oxforddictionaries.com/words/usage.
4 For a list of common Greek and Latin endings and their plural forms, see www.biomedicaleditor.com/spelling-tip-latin.html. A longer list of terms common in biology can be found at write-site.athabascau.ca/documentation/Irregular-plurals-Biology.pdf.
5 As a result of language change, in some fields it is becoming acceptable and even common practice to use data as a singular noun.
6 For a more detailed explanation and more examples, see www.scienceeditingexperts.com/using-respectively-correctly.
7 Their ontology.
8 More specifically, "Get in the car" in English is the same as "On car" (上車) in Chinese, as Chinese has no articles and "get in" is expressed by the single character 上 which means "on."
9 See, for example, www.ego4u.com/en/cram-up/grammar/prepositions.
10 For more details, see https://en.oxforddictionaries.com/usage/compare-with-or-compare-to.
11 Even more interesting is *identical.* The preferred form is identical to, but identical with is also found frequently in a corpus, and (unlike compare) seems to be more common with metaphorical usage. See https://lingohelp.me/preposition-after-adjective/identical-to-or-identical-with.
12 See https://en.oxforddictionaries.com/definition/different.
13 For a good summary of the debate, see www.dailywritingtips.com/try-to-vs-try-and.
14 This can also be done on a large group of documents at once (a corpus) using a concordance program. See Appendix 2.
15 Also known as Lake Webster.
16 Equipment is often considered countable and pluralized in Asian English, but not in American or British English.
17 www.oxforddictionaries.com/words/british-and-american-terms.
18 Wikipedia is not usually cited as a reference in academic work, but their article on the spelling differences between British and American English is actually quite a good summary: https://en.wikipedia.org/wiki/American_and_British_English_spelling_differences.
19 Unfortunately "Show all" is not on the menu by default in the current version of Word, but can be accessed by the "Tell me what you want to do" search. I find it useful to add it to the menu, using File/Options/Customize ribbon.
20 The sites will undoubtedly change rapidly, but at the time of publication include (alphabetically) Academic Karma, Paper Critic, Peerage of Science, Publons, Pub Med Commons, Pub Peer, and Rubiq. Particular services, requirements, and costs differ.

Reference

DAY, R. A. 1998. *How to write and publish a scientific paper,* Phoenix, AZ, Oryx Press.

Part 5

Adding the final touches

Chapter 24

Illustrating your manuscript

24.1 The purpose of graphics

In a speech to leaders of the communication industry, Rupert Murdoch said, "Great journalism will always attract readers. The words, pictures and graphics that are the stuff of journalism have to be brilliantly packaged; they must feed the mind and move the heart" (La Monica, 2009).

Whether in journalism or engineering, it is the combination of good writing and good graphics that attracts readers. Graphics is a collective term for all the tables and figures in an article.[1] The title and abstract may lead people to look at the graphics, but often it is the graphics in turn that will draw them to read the whole thing. To use an analogy, graphics are the 60-second trailer that entices people to sit down for the whole 2-hour show. Thus it is important not only to have graphics, but to ensure that they are high quality and tell the main story of your research, while leaving enough questions unanswered to make reading worthwhile.

> Like a preview, graphics should attract attention, showing some of the most interesting portions without giving away the whole story.

Graphics and text should supplement one another, without duplication of information. Graphics must be able to stand alone, so an experienced reader in your field can understand the main point of the graphic without looking at the text. At the same time they should complement the text so that both are required to get the full message, just as both the soundtrack and video are necessary to fully appreciate a movie. As you design your own graphics, consider the purpose of each. What does it add to the paper that makes it essential? Just like text, if it does not serve a purpose, eliminate it. Remember also that many journals now have supplemental materials online, providing another option for graphics of interest to a smaller audience.

Graphics, both tables and figures, can be used either to summarize or show detail. For example, a table may give the details, while the text summarizes the main point. Or a graph may show a trend, while the text explains possible causes of that trend. The basic principle behind choosing whether to express something using text, table, or figure is to consider what will be easiest for your reader to understand. Sometimes that basic principle trumps any rules presented in writing books.

The maxim "writing is an art" is true for all aspects of writing, but nowhere is it clearer than when producing graphics. The ideas presented in the following sections are not rigid rules, but general principles that may help you convey your ideas more effectively.

24.2 Terms used when referring to graphics

Before talking about the characteristics of a good graphic, I need to clarify terminology, because each kind of graphic uses a different set of terms and those for figures are different in British and American English. Table 24.1 and Figure 24.1 show the terms used for tables and figures, respectively, and the location for each. I will be using the American designations, in line with the rest of the textbook.

Note that the table title goes at the top.[2] This is easy to remember because all three words begin with T. On the other hand, the figure caption follows the illustration. The Writing

Table 24.1 Table terminology: title

	Column 1 heading (unit)	*Column 2 heading (unit)*
Row 1 heading (unit)	Data	Data
Row 2 heading (unit)	Data	Data
Summary row heading	Summary statistic	Summary statistic

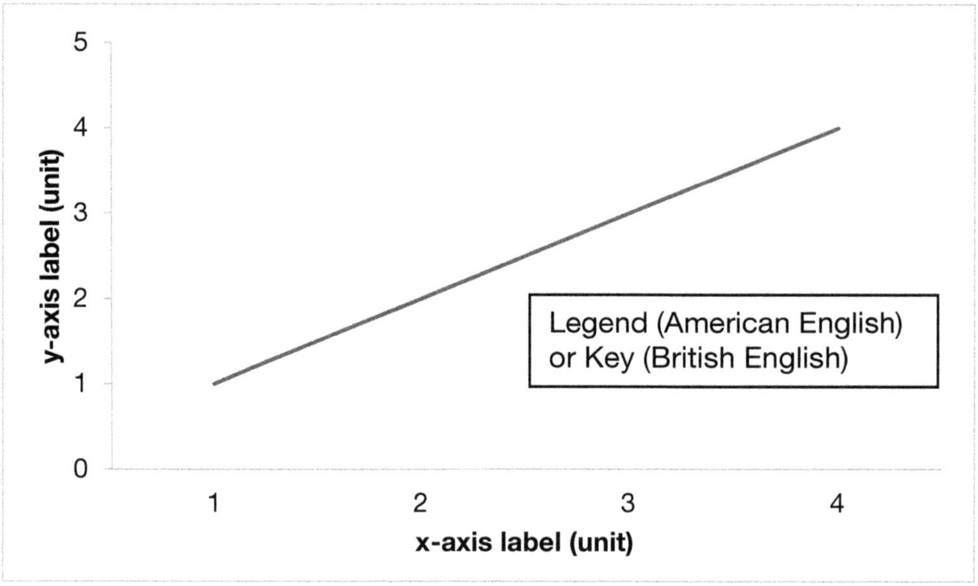

Figure 24.1 Figure terminology: caption (American English) or legend (British English).

Center at UNC-Chapel Hill helpfully suggests that this is because we tend to read a table from the top down, whereas we read a graph (the most common type of figure in science literature) from the bottom up, from the caption to the x-axis to the y-axis.[3] Note also that for column headings, the unit is often placed in parentheses on a separate line, as shown. For row headings or axis labels, the unit may be placed either in a separate line or after the name. While some writing books will state requirements for the placement of units, many journals now do not seem to care, and placement seems to be related to saving space rather than comprehensibility. Unless your journal lists specific requirements, think what would be easiest for your readers to understand.

> Titles go at the top of a table (Table Title Top, T^3), but captions go below a figure.
> Units are placed in parentheses below or after a heading or label.

24.3 Tables

Academic writing textbooks often state that tables should be used when the details are important, to show exact numbers and their precision, but sometimes fail to mention that tables are also commonly used to summarize data, whether numerical or non-numerical (categorical). For example, the authors of one recent excellent book on scientific writing (Cargill & O'Connor, 2013: 26) say that tables are

"most useful for

- recording data (raw or processed data);
- explaining calculations or showing components of calculated data;
- showing the actual data values and their precision; and
- allowing multiple comparisons between elements in many directions".

Immediately after that, the authors present a table, reproduced below as Table 24.2, summarizing the difference between tables and figures. It is ironic that this table does not perform any of the functions they just listed for a useful table! Nevertheless, it is useful as a summary of categorical differences in how tables and figures show numerical data.

Table 24.2 Example of a summary table showing non-numerical categorical data

The choice between tables or figures for data display.

Most useful	*Table*	*Figure*
When working with	number	shape
When concentrating on	individual data values	overall pattern
When accurate or precise values are	more important	less important

Redrawn from Cargill and O'Connor (2013: 26, Table 5.1), used by permission.

It is important to distinguish between tables that are primarily numerical and those that are primarily non-numerical, because there are differences in preferred layout based on the way our brains process information. In English we read words as whole units and normally read across a line of text, so it is more natural to compare words across a row. On the other hand, it is easier to compare numbers in a column, as we do not need to process the meaning of each number but can quickly compare their relative size based on place value.

It is easier to compare numbers in a column, but words written in a row, so numerical and non-numerical tables should be arranged differently.

The difference between the two arrangements can be seen by comparing Table 24.2 with Table 24.3. In Table 24.2, the categories the authors want us to compare are in columns, and we naturally read across each line to compare tables with figures. On the other hand, in Table 24.3, the categories to be compared are listed in rows and we can quickly compare the difference between the different arrays by scanning down each column of numbers, looking only at the position and value of the first numeral to compare their relative size.

You may have noticed that there are no vertical lines in either table. Traditionally, vertical dividing lines were not allowed because of difficulties in typesetting prior to computers, and some journals retain this restriction. Whether permitted or not, reducing the

Table 24.3 Example of a good numerical table

Table 1
Array size and number of stations

Name		*Smallest station separation (m)*	*Largest station separation (m)*	*Number of stations*
TCU062	*S* array	12.5	43.3	6
	M array	25.0	86.6	6
	L array	106.7	449.4	6
TAL001	*S* array	12.5	43.3	7
	M array	25.0	141.6	7
	L array	69.5	237.5	6
	X array	317.4	1192.1	6
TAL002	*S* array	10.0	34.9	7
	M array	19.8	61.2	7
	L array	84.0	227.6	7
TCU067	*S* array	21.7	40.4	5
	M array	35.0	84.0	6

No vertical lines.
Horizontal lines between:

- Table title/column headings
- Column headings/table body
- Table body/footnotes (or bottom)

Redrawn from Satoh et al. (2001), used by permission.

number of lines in a table makes it easier to read, just like eliminating any other unnecessary information. In many cases, differences in spacing between rows or columns can guide which way the table should be read and which material should be considered as a single block, even more effectively than lines could.

Horizontal lines are common between:

- Table title/column headings
- Column headings/table body
- [Table data/summary statistic: totals or means] (optional, or white space, as in Table 24.1)
- Table body/footnotes (or at the bottom of a table)

There are many details to consider when creating a table. First, check the instructions to authors to find the **requirements**:

- **Table size**: single or double column, or no restriction?
- **Font size**: minimum? (make sure everything will be readable as printed)

There are additional things you can do to make **numerical data** easier to read:

- **Necessity**: delete any data unrelated to the research goal.
- **Importance**: most important comparison first.
- **Order**: logical based on the most important comparison, not testing order.
- **Alignment**: align numerical data on the decimal, right align if no decimal.
- **Precision**: identical for all data of one type, only enough to allow comparison.
- **Summary**: report a mean or total to no more precision than the data.
- **Exponents**: place exponents in the unit rather than within the table.

There are often specific requirements for the **format of titles**:

- **Justification**: centered or left?
- **Placement**: title on the same line as the table number or the following line?
- **Font**: Arabic or Roman numeral, capitalization, bold, etc.?
- **Periods**: period after the number? Period at the end of the title?

Here are some further **suggestions for the title and headings**:

- **Short titles**: table titles in many fields of engineering are less than one line. Additional explanation may be placed as a caption at the bottom of the table.
- **Column headings**: center headings over data. By default, heading text is left justified and numbers are right justified, leading to misalignment.
- **No split words**: do not split words in headings. Make the column wide enough to fit the words, hyphenate by syllable, or abbreviate.
- **Units**: indicate units as a second line after the column headings, in parentheses, not in the table body.
- **Minimize white space**: eliminate extra space. Make each column wide enough for the headings and data, not wider, unless space is used to guide reading.

> Journals and theses have minimum requirements for table format, but you should do more, to make them easy to read.

The default settings for tables in MS Word do not produce a publication-quality table. By default, columns are the same width, text is left aligned, and cells have complete borders. Fortunately, it is easy to modify most of these settings. Since you want your paper to stand out for excellence, it is worth taking the time to find out how to use the features on the Tables tab. Here I will only note two procedures that are less well documented: aligning on a decimal and table footnotes.

To align on a decimal, open the ruler by clicking the box at the top of the left slider. Next, click on the tab marker until you see an inverted T with a dot next to it, meaning align on decimal. Then highlight the cells in the table that you want to align. Finally, click on the ruler where you want the decimal to be. You can adjust the position by sliding the marker.

Explanatory footnotes for your table should appear immediately below the table, not at the bottom of the page. In Word this can be accomplished by placing section marks before and after the table, using endnotes for the notes within the table, placing endnotes at the end of the section, and restarting numbering each section. Of course, this only works if you have no other endnotes in your document. Alternatively, you can use normal superscripts.

Exercise 24.1 Evaluating tables in your exemplars

(1) What types of tables are used in your exemplars (numerical or non-numerical)? Where are they (which division)? What is the purpose of each (what components are they associated with)?

Hint: Many will have two purposes. All summarize data patterns (Component 7), but many also use those data in a comparison (Component 8) or interpretation (Component 9).

(2) How long are the titles (number of words)? What is the format of the titles (justification, numbers, capitals or bold, periods)?

Hint: If your journal has an electronic template, the template should do the formatting automatically.

(3) Could any be improved?

Hint: Just because a graphic is published does not mean it is good. You need to fit in as far as the type of information presented but can stand out with appearance.

Teaching tip: Exercises 24.1, 24.2, and 24.3 expand on Exercise 8.1.

Teaching tip: Have students find poor examples from their exemplars and share them with the rest of the class, saying how they could be improved.

Teaching tip: Do not assume students understand how to use the features on the Tables tab in Word. Many do not.

24.4 Figures: graphs

Two distinct types of graphics are classified as figures in academic writing: graphs and illustrations. Most of the discussion in writing books is about the former, as illustrated by Table 24.1, in which "figure" means a graph displaying the same data as the table. Illustrations usually are mentioned only in passing, almost a footnote. Nevertheless, since illustrations are important in engineering, I will discuss them separately in the next section.

Graphs are common in the Results or Testing division as a way of summarizing data. Line graphs are used to show change over time or across a range of values of a continuous variable. Bar graphs are required when the variable is discontinuous. Pie graphs and radar graphs are rare in science and engineering.

> Do not be satisfied with a default graph, which almost always contains unnecessary or poorly ordered output.

As recently as 30 years ago, graphs were drawn by hand by an author and reproduced by graphic artists. Every unimportant point or line was eliminated. Now graphs can be generated by a computer with almost no human intervention, but as a result they often contain far too much information, making them hard to read.

Here are some ways to make it easier to read your **graphs**:

- **Highlight**: most prominent line = heavy line weight, dark solid color.
- **Importance**: highlight the current standard, or your work?
- **Simplify**: not too many lines on one graph.
- **Scale**: same scale on similar graphs facilitates comparison.
- **Comparison**: emphasize similarity or difference by choice of scale.
- **Axes**: time or experimental variable = x-axis; response variable = y-axis.
- **Error bars**: only if necessary; consider showing fewer, or a cone of variance.
- **Color**: still distinct if printed in black and white?
- **Boxed**: few points = box around graph; complex = no box.

Many of the points mentioned with reference to table titles apply to figure captions as well. The following are more specific to **figure captions**:

- **Captions**: often longer and more descriptive than titles, sometimes sentences rather than phrases; may contain explanation of axes or units in a second sentence.
- **Axis labels**: begin with a capital letter, but following words in small letters.
- **Units**: in parentheses, following label.

- **Consistency**: maintain order of lines and patterns for all similar graphs.
- **Legend**: order of legend should match order of lines on the graph, top to bottom.

Exercise 24.2 Evaluating graphs in your exemplars

(1) What types of graphs are used in your exemplars (line, bar, or other)? Where are they (division)? What is the purpose of each (component)?

Hint: Many will have two purposes (see Exercise 24.1).

(2) Are the captions phrases or sentences? How long are they? What is the format?
(3) Could any be improved?

Hint: Discuss your suggestions with a partner. Different people will have different ideas on what is good or could be improved.

24.5 Figures: illustrations

In sharp contrast to science articles, illustrations are by far the most common type of graphic in engineering, in many journals surpassing the number of tables and graphs combined. As mentioned in Chapter 8, in IPTC they frequently are found in the Process division, or even Introduction. In some cases they may depict a physical arrangement, either schematically or photographically, that is hard to describe in text. They may also represent an abstraction such as a waveform. Occasionally they show a flowchart or other process. As with graphs, one of the most common problems is too much information.

> Illustrations are far more common in engineering than science, frequently depicting some aspect of the current design.

Illustrations are very field-specific, so it is hard to give general guidelines. What would be considered too cluttered in one field might be considered just right in another. Some journals prefer a smaller number of figures, even if that means several very loosely related figures are grouped together.

Nevertheless, some general suggestions can be made about **illustrations**:

- **Necessity**: remove unnecessary or repetitive information.
- **Photographs**: retain proper aspect ratio (length to width) when resizing.
- **Background**: declutter photographic images by cropping.
- **Resolution**: clear, generally minimum 300 dpi.
- **Text**: large enough to read in the final print version, not on your monitor.

Exercise 24.3 Evaluating illustrations in your exemplars

(1) What types of illustrations are used in your exemplars (photograph, diagram, schematic, other)? Where are they? What is the purpose of each?

Hint: Illustrations are frequently used to depict details of the current study, but some show results or comparisons.

Hint: You may want to consider different types separately.

(2) Are the captions phrases or sentences? How long are they? What is the format?
(3) Could any be improved?

24.6 Graphics that break the rules

Many books on writing present a list of rules for what should be included in a table, what should be presented in a figure, and what should be mentioned only in the text. Unfortunately, such rules are made to be broken. The main consideration is what will be easiest for the reader.

For example, some authors say that a list of conditions should be reported only in text. Yet Table 24.4 shows an example of a type of table that is very common in engineering. Since it simply lists the operating conditions of the system and could easily be presented as a single sentence of text, we need to ask why it is so common. The obvious explanation is that it is easier both to find and to read, just like bullet points in a conference presentation.

Sometimes rules are broken for the sake of ease for the author rather than the reader. For example, in engineering it is common to compare several dozen characteristics of two to four designs. Ideally, according to what I said above, each characteristic should be one column, so readers can compare similar results vertically. Yet in many papers characteristics are listed as rows. This saves space and formatting time, since fitting the names of all those characteristics (sometimes several words) is easier in row headings than column headings. As long as the number of designs compared is not too great, this seems to be acceptable.

For any rule that is given, including the suggestions I made above, there will be exceptions. If a certain type of graphic is common in your field, feel free to use it even if it seems to go against rules set by writing instructors who do not know your field. You

Table 24.4 Sample table of operating conditions, common in engineering

TABLE I
SPECIFICATIONS OF THE PV ARRAY

PV array	Rated power	900W
	Open circuit voltage	150V
	Short circuit current	7A
	Connection	6 (each 150W)

> The only two inviolable rules are that each graphic must be mentioned in the text and each should be helpful to the reader.

need to fit in. On the other hand, think about whether it really is a useful way of presenting the information or if there might be a better way. Even if journals in your field publish tables using the MS Word default showing all the gridlines, could you make your graphics more visually appealing? You want to stand out for excellence. The only truly inviolable rule is that every graphic must be mentioned in the text.

Evaluate your understanding

(1) How do table titles and figure captions differ?
(2) How should the arrangement differ for numerical and non-numerical tables?
(3) Why is it beneficial to minimize the number of horizontal and vertical lines in a table?
(4) Why should the instructions to authors be considered minimum requirements rather than the goal?
(5) How can you improve the default output of your graphing program?
(6) Why are illustrations more common in engineering than science?
(7) Why is it impossible for a writing textbook to give absolute rules on what makes a good graphic?

Homework

(1) Format your graphics according to the instructions to authors for your journal. Consider how to improve them using the suggestions in this chapter.

Hint: Just like your text, graphics should be revised several times. Do not simply accept the preset output format from whatever program or device you are using.

Teaching tip: Students doing peer review sometimes forget that standards differ by field. Encourage them to ask their partners why they chose to present a graphic a certain way or use modal-based questions (Would it be possible . . .?, Could you . . .?) to make their suggestions more polite.

Supplemental material:

Class exercise 24.1: Evaluating sample tables

Class exercise 24.2: Evaluating sample graphs

Class exercise 24.3: Evaluating sample illustrations

Notes

1 As mentioned in section 8.4, students using English as a lingua franca may have trouble distinguishing graph from graphic, but there does not seem to be a better general term.
2 This is almost universal, but I have seen one computer science journal (*Journal of Machine Learning Research*) where both table titles and figure captions are placed under the graphic.
3 http://writingcenter.unc.edu/handouts/figures-and-charts.

References

CARGILL, M. & O'CONNOR, P. 2013. *Writing scientific research articles: Strategy and steps*, New York, John Wiley & Sons.

LA MONICA, P. R. 2009. *Inside Rupert's brain*, London, Penguin.

SATOH, T., KAWASE, H., IWATA, T., HIGASHI, S., SATO, T., IRIKURA, K. & HUANG, H.-C. 2001. S-wave velocity structure of the Taichung basin, Taiwan, estimated from array and single-station records of microtremors. *Bulletin of the Seismological Society of America*, 91, 1267–1282.

Chapter 25

References and citations

25.1 When the writing is done

I suggested in section 20.3 that you use placeholders for citations while writing and postpone adding the actual citations and references until the end. This allows you to become more familiar with the authors and dates of important papers in your field and to focus on writing instead of bookkeeping details, while avoiding technical glitches.

In this chapter I will first talk about two different reference and citation formats. This will be followed by a discussion of the benefit of using reference managers to avoid excessive retyping and reformatting.

25.2 Reference and citation formats

There are two basic citation-reference formats, each with many different variations, the numerical system common in engineering and the author-year system common in science. Recall that *citation* refers to the marker in the body of the article that directs readers to the *reference* listed at the end.

The easiest way to choose the right format for your journal is to use one of the reference managers described in the next section. This section simply provides a little background so you can appreciate how much those tools do for you, and things to check to make sure that your reference manager has actually done the formatting correctly.

> The two basic formats are numerical and author-year, but the many minor variations from journal to journal make it very worthwhile to use a reference manager.

Almost all engineering journals use some variant of the numerical system, in which citations are indicated by a number in the text. The citation number may be shown in brackets, parentheses, as a superscript, or in superscripted brackets, and thus will appear as [1], (1), ^{1} or $^{[1]}$, respectively. Multiple papers within a single citation are usually listed in numerical order, e.g., [1, 2]. The citation marker may occur within the sentence or after the punctuation.

There is even more variation when it comes to the references. References may be listed numerically in order of appearance in the article (citation order) or alphabetically by family name[1] of the first author (reference order). The date of publication appears near the end of the reference. Depending on the journal, either full names or initials may be used for authors' given names, and either given names or family names may be listed first. If initials are used, some journals require use of a period after the initial while others forbid it. All authors (or at least ten) are listed. References are often but not always listed in block style, with the number to the left and space between each reference. This list of differences is not exhaustive, but gives some idea of the amount of variation from one journal to another.

Still more variation exists in the author-year format common in science journals, because of the variation in citations. The basic format for the citation is the family name and date of publication placed in parentheses in the text, e.g., (Alpha 1993). Sometimes the two are separated by a comma, but not always. If there is a second author, that author's family name is added, with either "and" or "&"[2] between the two names, e.g., (Alpha and Beta 1998) or (Alpha & Beta 1998). Sometimes a third author's name may be added, but usually additional authors are indicated by "et al."[3] If multiple papers are mentioned in a citation, some journals require that they be listed in alphabetical order, and others by date of publication, e.g., (Alpha 1993, Beta 1990) or (Beta 1990, Alpha 1993). Citations are usually within the sentence, before the period, but not always.

The variations in the references in author-year format are similar to those listed for the numerical system above. The main differences between the two systems are that the year of publication is placed immediately after the authors' names (always listed family name first) and a hanging indent is common, often without space between entries.

Exercise 25.1 Entering references and citations

(1) Examine the citations and references in several different journals. List the differences between them. What system and which variations are used in the journal you are writing for?

Hint: The instructions to authors for your journal will contain specifics on the format they require, along with examples for various types of sources.

> *Teaching tip: Instead of trying to teach the differences, allow students to compare the format of references in several different journals, noting as many differences as they can. See Class exercise 25.1.*

25.3 Reference management

Less than a generation ago authors would spend hours typing references in the format required by each journal, but those days are fortunately behind us. Instead, the question now is which of the multitude of reference management tools to use. Since the tools available are changing quickly, I will only comment on the advantages

> Reference managers are changing rapidly, with many younger researchers turning to cloud-based platforms.

and disadvantages of different types of tools and some of the better known examples rather than all of the individual programs. There are currently three types of reference management tools available—those built into word-processing software, separate reference managers that run on your computer, and online cloud-based accounts.

Word-processing software has the advantage of integration. Since the reference manager is an integral part of the program, compatibility is not an issue. Those writing in LaTeX will want to use the integrated BiblioTeX reference manager. In MS Word the main disadvantage is the limited number of formats supported. Initially only three were offered—APA, Chicago, and MLA—which was not very useful for scientists and engineers. Recently several others have been added, including IEEE, and the choices will undoubtedly expand in the future, but for now it is still quite limited. Currently sources must be typed or copied into the master list manually, another severe limitation compared to the other two types.

Computer-based reference managers, such as Endnote, support formats for thousands of journals.[4] Just as important, it is possible to download references from many online sites directly into the database, eliminating the need to type in the information. On the negative side, the software is not free, takes space on your disk, and your database may be lost if your computer crashes. Some schools or libraries offer a site license on one of the reference managers so faculty and students can use it without paying individually, but that license expires if you move to a different school without such an agreement, requiring you to purchase an individual license or export your data to a different program.

Online reference managers are the latest addition to the market, and the choice of many younger researchers who grew up on computers. Some of the larger, better known sites include Mendeley, RefWorks, and Zotero, but there are many others as well, catering to various needs. These also allow direct import of references from journals or search engines and have many export formats available. If you are not familiar with reference managers, I suggest you ask your colleagues what program they are using, as they will be the best resource on current options.

No matter what management system you choose, the input method is similar. Sources are added to the master list, those needed for a particular paper are selected from that master list, marks are placed in the text to indicate the location of citations, and the style of output is set. The program does the rest. Well, not quite.

> It is still up to you to do the final check to ensure that the reference manager has output the results correctly.

It is still incumbent on the researcher to check the printed references. *Errors in the reference list usually reflect errors in the database* that need to be corrected in the master source list. Odd sources, such as articles in an edited volume that is part of a series, can be a particular problem, and may require trial and error to find out how to enter them so they display correctly.

Evaluate your understanding

(1) What are the two basic types of citation/reference formats, and some of the variations within each?
(2) What types of reference managers are available, and what are the advantages and disadvantages of each?
(3) How can errors in references arise, even when using a reference manager?

Homework

(1) Using whatever method you choose, add all the necessary references and citations to your document in the proper format.

Teaching tip: Prepare a sample document with a few references. Show how it is possible with a few mouse clicks to change the format of both citations and references. Show the number of formats available in other reference managers compared to Word.

Teaching tip: This chapter and the following chapter may be taught in one class, as none of the exercises takes long to explain or complete.

Supplemental material: Class exercise 25.1—Differences in reference formats.

Notes

1 I use "family name" and "given name" instead of the more common American designations "last name" and "first name" because in many Asian languages the family name is written first.
2 This mark is called an ampersand, derived from the Latin word *et*, meaning "and."
3 The abbreviation "et al." stands for the Latin *et aliquid*, meaning "and the rest." Since *et* is a full word, but *aliquid* is abbreviated al., there is a period only after the latter.
4 More common formats are included in the Endnote software; others can be downloaded from their website.

Chapter 26

The first shall be last

26.1 Attracting attention

The title and abstract are the first parts of your paper that a reader will see, but the last parts written, a new twist on a classic paradox. After completing a 10,000-word article, writing another 300 words might seem like a small task, but these parts are vitally important to the success of your paper and should be taken seriously. If poorly done, all your hard work in the paper will be for naught, because these final bits you write are what attract readers. If graphics are the 60-second trailer that gets people hooked, the title and abstract are the 30-second preview that catches their attention in the first place.[1]

I can't vouch for the accuracy of the numbers, but it has been estimated that the number of people reading an article follows a law of powers of 10. Of those who read the title, only 10% will read the abstract. Of those who read the abstract, only 10% will skim the article, looking at the graphics. Of those, only 10% will read the whole article carefully. If this is true, only one out of a thousand people who read your title will read the whole article. The numbers are certainly not exact, but the principle holds true—if you want people to read the article, they have to find it, and find it interesting. I will cover the various attention-catching parts in the order I suggest you write them.

26.2 Abstract

Although you may have written a preliminary abstract before beginning to write,[2] you will need to completely rewrite it once the paper is complete. Not only will your thoughts be clearer now, but you have undoubtedly made changes in your aim, procedures, or conclusions in the course of writing. You may have included things you did not plan, while eliminating others.

To understand the structure of an abstract, let us return one more time to our exemplar articles. The abstract is a summary of the entire article, thus it frequently contains something from each of the divisions, often in the same order but not necessarily the same proportion as the article.

Although in science the focus of the abstract tends to mirror the focus of the articles themselves, with much of the abstract devoted to the Results and Discussion, this is not necessarily the case in engineering. In IPTC more attention may be focused on

In science most of the abstract is devoted to Results and Discussion, just like the article, but in engineering little space may be devoted to the new Product or Process.

either the need for the work or the contribution rather than the design itself, perhaps because it is hard to briefly summarize so many design details. Half of the abstract may be devoted to the Introduction and Conclusion.

To demonstrate, let's return to our sample article, "Spatial query integrity" (Hu et al., 2013). Table 26.1 shows key phrases that allow us to recognize the components represented by each sentence, and the divisions where each component is found. Half of this abstract is related to the Introduction, defining the problem and stating the proposed solution. On the other hand, only one sentence summarizes the Process division, and even that refers to the framework rather than the design details, which comprise more than half of the article. This is followed by a two-sentence summary of the testing, including how the new system compares to the current state-of-the-art solution, and therefore the contribution of the new design.

It is also important to pay attention to the length of the abstract. In some journals, the abstract may be as short as 4 sentences or 100 words. In others it may be 12 or more sentences and over 300 words. Even longer abstracts are common for theses. Conference abstracts vary considerably. Some journals with medical applications have a required abstract structure that allows practitioners to quickly find the information they need.[3] Needless to say, your abstract should match the requirements of your target.

Exercise 26.1 Structure of exemplar abstracts

(1) How many sentences do the abstracts of your exemplars contain? What is the average number of words per sentence?

Hint: Use exemplars of whatever genre you are writing, if available.

Table 26.1 Components in each sentence of the abstract of "Spatial query integrity"

#	*Component markers*	*Comp.*	*Div.*
1	... has grown rapidly over the past few years.	1a	I
2	... has provided a flexible and cost-effective platform ...	2b	I
3	Nevertheless, ... remains a challenging problem.	2b	I
4	In this paper, we focus on ... and propose an efficient scheme ...	3b	I
5	Our approach is based on ...	4d	P
6	We evaluated ... on real-world datasets using mobile devices ...	6c	T
7a	Compared to the current state-of-the-art ...	8d	T
7b	... produces significantly smaller verification objects and is more computationally efficient ...	10b	C

(2) Based on your knowledge of the components, assign each sentence to a component and to the division where that component usually occurs in an article. Do the sentences appear in the same order as the order of the divisions? How many sentences are devoted to each division?

Hint: Sometimes a sentence may include more than one component, for example results (7) and comparison (8).

(3) Is there a required structure, as in some medical journals?

Hint: This is relatively uncommon.

Hint: Could you present all the information in this Exercise in one table, or would it be better to create more than one, or should some be stated only in the text?

> *Teaching tip: It is very worthwhile to have students compare the tables they create for the abstract. With so many different types of data, there will be many options, and it will help them evaluate what works well and what does not.*

Now let's consider how to write your own abstract. An abstract will almost always include a statement of the research goal, and something about the framework to indicate the connection to previous research. It will also include a summary of any comparisons made, whether to expectations or previous research. It may include an introduction of either the background or importance and a one-sentence conclusion discussing implications, applications, or limitations of the study.

Present tense and active voice are common in abstracts. While it would be possible to use the past tense to describe the research you did ("in this study we proposed a new solution"), it is more common to use the present tense to describe the paper that immediately follows ("in this paper we propose a new solution"). Similarly, although it would be possible to write in the passive voice ("in this paper a new solution is proposed"), it is usually clearer and more concise to write in the active voice.

> Copy ideas but not whole sentences from your paper, then rewrite for conciseness, coherence, and clarity.

Students often wonder whether they can copy sentences from the body of the paper into the abstract. While this does not constitute plagiarism, it is generally not a good idea, as the abstract is much more concise than the body of the paper.

I recommend following the 7Cs of Change when writing your final abstract. Begin by selecting important sentences from your paper and copying them into a new document. Next eliminate any unnecessary or unimportant words or phrases (conciseness). Then rewrite the remaining phrases as new sentences to create a logical flow of thought (coherence). Finally, consider how to link the sentences together or modify your claims to make your argument clearer and stronger (connection and connotation). Particularly for beginning writers, this ensures that all the main ideas from your paper are present, but nothing else.

There are different types of abstracts, including conference abstracts and the abstract of your thesis in addition to the abstract of an article. Each is slightly different, but in some situations parts of one can be reused for other purposes, as we will see in section 27.6.

Exercise 26.2 Writing the final version of your abstract

(1) Using the information on what is expected in an abstract in your field from Exercise 26.1, rewrite your abstract.

Hint: The key arguments in your article should be clear from the abstract alone. Do not simply narrate what you did, but claim and support that your work makes an important contribution.

26.3 Title

The title is the only part of your document that most people will see. Its purpose is to allow researchers in the field to decide how relevant your work is to theirs. Thus a good title will be self-explanatory to your target audience, containing words that identify both the purpose and contribution of your research.

Research article titles may be noun phrases, statements, or questions. Noun phrases are by far the most common. Statements may be used in science in situations where a simple answer to a research question can be stated (Cargill & O'Connor, 2013: 66). Questions are far more common in the social sciences, where the research goal may be framed as a question. The length and amount of information contained vary greatly by field. While titles of fewer than 10 words are common in electrical engineering, titles may be up to 30 words in some chemical engineering journals. Colons are common in some fields, and may indicate various relationships, including:

- Topic: Method;
- Problem: Solution;
- General: Specific;
- Major: Minor.

> **It may be possible to turn your research goal into a title, with the contribution first to foreground it, followed by the general research area.**

Titles often indicate both the general topic of the research and what makes the paper unique—its contribution. One of the easiest ways to make sure your title summarizes your article is to turn your research goal into a title. Try to make the key contribution the first part of the title, with the research area second, to foreground what is unique about the current research. The two parts of the title are usually linked with a word that indicates the nature of the connection, for example *with*, *for*, *by*, *using*, or *based on*. Your title may also include words that identify the nature of the study, such as preliminary, case study, framework, or simulation. While it is important to be concise in all scientific writing, it is even more important in the title, where the importance of every word should be weighed.

Exercise 26.3 Creating a title for your article

(1) How long are the titles of your exemplars? Are the titles statements or questions (containing a verb) or noun phrases (no verb)? Do any contain a colon? If so, what is the relationship between the first and second parts?
(2) Do the titles include the research goal or main contribution of the research, the general topic or research area, or anything else? In what order?

Hint: To get a broader sample, examine the titles of all the articles in your references.

(3) Write several different versions of a title before choosing the best.

Hint: Try to be as concise as possible without losing clarity, placing the most important information near the front.

26.4 Keywords

In order to choose good keywords, it is essential to understand how they are used, which requires a little historical context. Keywords, also called key words or index terms, are a holdover from the pre-Internet era, when they were used to compile printed indices to help people search for articles. When I was a graduate student, the only way to find articles was to go to the library and pore over the small print in those huge volumes. Since titles often convey only the specific topic, keywords were used to place an article within its wider context, allowing more people to find it. With the rapid rise in computing power, modern search engines search not only the title and keywords, not even just the words in the abstract, but all the words in the article. Thus keywords no longer fulfill the purpose they were created for. Now they are more likely to be scanned visually by a researcher, along with the title and abstract, as part of the decision whether to read the article or not. Thus *there is little or no benefit to listing words in your title as keywords.*

> Choosing keywords not in your title can help extend the appeal of your article to a different audience.

Some journals require or suggest that you choose from a predetermined list of keywords or index terms.[4] Where possible, specific words in your title should be supplemented by keywords representing the broader field of study, without duplication. On the other hand, depending on how many keywords the journal requires, it may be better to duplicate words in your title rather than choosing words that do not fit.

Exercise 26.4 Deciding on keywords

(1) Check the instructions to authors to find how many keywords are required, and whether they must be selected from a list.
(2) Look at your exemplar articles to see which keywords were chosen by others studying a similar topic.
(3) Choose suitable keywords for your article.

Hint: Think outside the box. Are there certain techniques used in your article that did not make it into the title, or perhaps an application?

26.5 Highlights

A few years ago Elsevier, a major academic publisher, began requiring highlights for all journals they publish, adding yet another category for authors to write. Highlights are intermediate in length between the title and abstract.

According to the Elsevier website:[5]

> Highlights are a short collection of bullet points that convey the core findings and provide readers with a quick textual overview of the article. These three to five bullet points describe the essence of the research (e.g. results or conclusions) and highlight what is distinctive about it.

Highlights should list the core results of the article, rather than importance, and may thus have a different focus than the abstract.

Thus the author is asked to write three to five highlights, each no more than 85 characters including spaces, listing the core results of the article. By asking for only the core results, other statements that might be in the abstract, such as background and application, are specifically excluded.

A quick analysis of their examples shows that each highlight is a simple declarative sentence containing about ten words, including one verb in the present or past active form. The content depends on the article, but generally includes material that would be found in the Results and Discussion or Process and Testing divisions. They may include the material studied or method used, but no details.

Exercise 26.5 Writing highlights

(1) Check the instructions to authors to find out whether highlights are required. If they are, look at examples in the online version of your journal. Do they fit the general pattern suggested above for length and sentence structure? What divisions of the paper are represented?
(2) Write your own highlights following the same pattern.

Hint: As with the title, write several versions before deciding on the best.

26.6 Acknowledgments, author information, and other bits

Depending on the genre, there may be various other short parts to be written before submission. These may include acknowledgments, author information, and a note of online material.

Funding sources or previous publication in conference proceedings should be noted in the acknowledgments.

If you (or your advisor) received a grant that paid for any part of the research (including your graduate assistantship), you must acknowledge it. There is a standard format for this that can be copied verbatim from other journal articles, apart from the grant source and number.[6] If the material in the article was previously presented at a conference, this should also be noted, along with previous publication of any portion of the material. Occasionally authors will acknowledge the assistance of someone who is not an author. This is less common in engineering articles, which tend to have more authors, than science, where technicians may be recognized as having done some of the labor although not contributing to the writing.

Some publications print information on the authors' affiliations and contact information, while others require a brief bio. These may be part of the article template provided by a journal or may be a separate section of the submission process if a template is not provided. Similarly, there is a standard procedure for referring the reader to online material, if it exists.

Exercise 26.6 Gathering the last bits

Write the final parts, following the requirements of the journal, department, conference, or funding agency.

Hint: You may need to gather information from your advisor or other authors.

Evaluate your understanding

(1) Why is it essential to rewrite your abstract after the paper is complete?
(2) Why do abstracts in science tend to reflect the structure of the article, while those in engineering may not?
(3) What principles can help you create a better title?
(4) Should keywords duplicate words in the title?
(5) How do highlights differ from the abstract?
(6) What types of information should go in the acknowledgments?

Homework

(1) Ask a colleague to comment on the clarity of your abstract, and which of several possible titles sounds best or most interesting.

Hint: Ask someone not in your lab, as others in your lab will be too familiar with the terms and procedures used to evaluate the writing as an outsider would.

Notes

1 The general pattern of researchers scanning the title and author, then abstract, followed by graphics when deciding what to read has been documented by Berkenkotter and Huckin (2016: 3), although they note that it differs somewhat depending on how central an article is to the reader's research area.
2 Chapter 3 homework.
3 For example, an IMRD oncology journal requires the abstract to discuss separately the Background, Methods, Findings, Interpretation, and Funding.
4 Including all those published by IEEE.
5 www.elsevier.com/authors/journal-authors/highlights.
6 This work was supported [in part] by [grant agency] [grant number].

References

BERKENKOTTER, C. & HUCKIN, T. N. 2016. *Genre knowledge in disciplinary communication: Cognition/culture/power*, London, Routledge.
CARGILL, M. & O'CONNOR, P. 2013. *Writing scientific research articles: Strategy and steps*, New York, John Wiley & Sons.
HU, L., KU, W.-S., BAKIRAS, S. & SHAHABI, C. 2013. Spatial query integrity with Voronoi neighbors. *IEEE Transactions on Knowledge and Data Engineering*, 25, 863–876.

Chapter 27

Submission and review

27.1 Warning: not every article is accepted (but it's not the end of your career)

Papers written for a class are accepted. What grade you get depends on the quality of your work, but acceptance is not an issue. Many conferences also accept almost every submission, but when it comes to more prestigious conferences or journals, perhaps the submission site should come with a warning label—Warning: not every article submitted will be accepted. Like most warning labels, it tells us something we already know, or should know, but don't want to think about. On the other hand, perhaps it should also include an encouragement—but a rejection is not the end of your career as a researcher. Almost every author has experienced rejection at some point.

In this final chapter I will focus on submission of a research article, but the same principles apply to conferences, grants, and other academic writing. There are four possible fates for every document submitted—acceptance, conditional acceptance, request for revision, and rejection. Outright acceptance without change is extremely rare. The best an author can realistically hope for is conditional acceptance: a promise of acceptance if certain concerns are addressed or changes made to the satisfaction of the editor. Requests for revision are quite common, where the author is asked to make certain changes, then resubmit, and acceptance or rejection is dependent on those changes. The rejection rate depends on the journal, with more prestigious journals having a higher rejection rate.

Getting a rejection does not necessarily mean your paper is bad or unpublishable. Frequently the problem is that it does not meet the requirements of that journal. I edited one paper for a seismology professor twice, but the editor rejected it twice, saying each time that it needed to be edited by a native speaker of English, without mentioning any specific problems. Although the grammar and spelling were correct, perhaps the word choice was different from other articles in the journal, or perhaps the argument structure did not match what the editor wanted. At my suggestion, the professor submitted the same article to a different journal and it was accepted the first time, with only minor revisions.

27.2 Choosing the right journal

Submitting your paper to the right journal will greatly increase your chance of acceptance, while submitting to the wrong journal dooms your effort. Two factors are important: the type of papers the journal accepts and the quality of your research relative to the impact factor of the journal.

It is easy these days to find out what type of papers a journal accepts, by going to the homepage for the journal and looking for one of the following, or some related phrase:

- Instructions to authors;
- Author guidelines;
- Guidelines for manuscript submission;
- Submission requirements.

Somewhere on that page is likely to be a statement of "Scope" or "What types of papers we accept." For example, the *IEEE Transactions on Power Electronics* Guidelines for Manuscript Submission include the following:[1]

> IEEE Transactions on Power Electronics' scope includes new results in the field of power electronics. For example, papers that treat original component, device, module, circuit, control, system, or application issues are of interest. Papers that do not show sufficient overlap with interest in the power electronics field (for example, papers on topics such as device physics, component manufacturing technologies, and circuit theory) should be submitted to other IEEE Transactions that cover these other fields. Authors should pay particular attention to documenting their work in relationship to the known literature. Papers of a historical or tutorial nature within the scope of this publication will also be considered.

This statement of scope makes it clear that certain papers marginally related to power electronics are not likely to be accepted. It also gives a hint that in order to be accepted, authors need to cite papers that document the connection between the current work and past work published in that journal. This is a good reminder, and true for all journals, whether specifically stated or not.

The impact factor of a journal is harder to find. Journals that are highly rated tend to publicize their high impact factor, but those are not likely to be the journals you will submit your first paper to, unless you happen to be lucky enough to discover something as important as the structure of DNA.[2] Impact factors measure how frequently articles in that journal are cited, which does not directly translate into importance, as some fields tend to have higher numbers of citations per article than others, which automatically boosts the impact factor of every journal in that field. Complicating the matter, there are now several competing systems for rating journals.

Unfortunately, there is sometimes a tradeoff in journal choice. The scope and readership of a journal is the most important factor in getting your work read by the target audience. Thus smaller, more specialized journals may give you better access to researchers in your specialty who are likely to read your work. On the other hand, the promotion process often gives extra weight to papers published in journals with a higher impact factor. Thus, you may eventually want to publish in both—the first to get your name known in the field, the second to look good on your resume or promotion application.

Many factors influence choice of journal, including scope of the journal, impact factor, and stage of career.

If you are relatively new in your field, the best way to determine which journal will be most suitable is to ask your advisor. He or she will know the journals in your field and what type and quality of papers each accepts, and thus which is most likely to be willing to accept your work.

27.3 Submission methods

Submission methods have changed drastically over the last two decades, and will undoubtedly continue to change rapidly with future advances in technology, so in this section I will focus more on basic principles rather than actual methods. Twenty years ago the researcher had to print and send three complete copies of the manuscript by what is now known as snail mail (in those days it was just mail) to the editor, who then forwarded one copy to each reviewer. Mailing back and forth took time, so an author would often hear nothing for months. Email submission became common as researchers got connected to the web and email systems were developed that could handle attachments. Now almost every journal has some sort of electronic submission system to standardize the process, although it may still take just as long to get a response. Nevertheless, the basic principles of submission remain the same.

Before beginning the actual submission, it is important to find out what will be required. In some cases the entire article must be uploaded, print-ready, on a template provided by the journal. For others, different sections of the article must be uploaded separately in different blocks on the submission form or as separate files. For example, the title, abstract, and keywords may have to fit within blocks of a certain length to ensure they meet length requirements, or tables and figures may have to be attached separately. Some journals require items that are not part of the article itself, such as author bios or a cover letter. You may also have to provide affiliations, physical and email addresses, phone numbers, or other information on the authors, and should have those available in electronic format before you begin the actual submission process. Some forms may require phone numbers or addresses in a format that does not match those used in your area, and you will have to find a way around this. This problem seems to be more common in submission to a conference than to a journal. If files must be uploaded, make sure you check the requirements for file type, size, and naming requirements. I suggest that you gather all the documents you need for each submission in one folder, so it is easy to find and access them while doing the submission, and months later after you get a response.

> Before beginning an online submission, look through what will be required and gather all the materials you will need in the proper format.

If you are asked to provide a cover letter in an electronic submission, do not duplicate information found elsewhere on the electronic form such as author names and institutions, the journal you are submitting to, or a statement that the research is original and not submitted to or published in any other journal, as was common in hard copy and email submission. What can be included are clarifications of anything that might be unclear, such as any problems you had entering data on the electronic form, or mention of supplemental material.

Other things that should not be included in the cover letter include accomplishments of the authors or advocacy for the paper. The paper should speak for itself. It is also not necessary to include a request for a response, invitation for the editor to revise, or indication of a willingness to revise. The editor knows the expectations better than you do. Any additional comments of this nature will undoubtedly be interpreted negatively, and should be avoided.

If separate files are required, make it easy for your editor. Do not use the authors' names, as review is often intended to be blind, but rather give a short but distinctive version of the title. If there are additional files, such as graphics, give them the same title followed by a part designator, e.g., Article title Fig 1.

Exercise 27.1 Preparing for submission

Look through the submission process for your journal. What files will you need, in what format? What additional information will you need?

Hint: Gather all the information you need for the submission in one folder. Once you have everything ready, submit—but not before checking everything one last time with all your coauthors.

27.4 How the review process works

Wolfgang Pauli, the famous theoretical physicist, stated, "I don't mind your thinking slowly; I mind your publishing faster than you think" (in Alley, 1996: 178). The purpose of review is to ensure that you have thought through everything before you publish, and that your work is of high enough importance and quality to merit publication. Reviewers may point out weakness in citations, description of the current research, data, or analysis—in other words, all the elements of evidence and inference we have talked about.

After a paper is received by a journal, the editor will usually take a quick look to make sure the submission is complete and within the general scope of the journal, then send the submission to at least two reviewers. They read the paper, answer a set of questions, write comments, and often give an overall assessment of the quality of the paper or a recommendation on whether to accept or reject. After they complete their review, the editor will summarize their feedback and comments, make a decision, and send the results to you.

First-time authors often wonder what the editor and reviewers are looking for. The types of questions reviewers evaluate could be summarized in three groups, leading to four different fates, as shown in Figure 27.1. This is obviously an oversimplification, as many different factors affect the final decision, including the personal preferences of the editor, but gives a rough approximation of the process.

Table 27.1 shows examples of the types of questions in each group. Although the specific questions vary, they tend to fall into the same categories.[3] As shown in the last two columns, all the questions are related to topics covered in this text.

The questions are listed in general order of importance, so outright rejection may occur for failure of any question in the first group. A paper may be rejected if the topic is considered unimportant or the solution not novel enough to merit publication, or if the argument is irreparably flawed or incomprehensible, making it impossible to assess

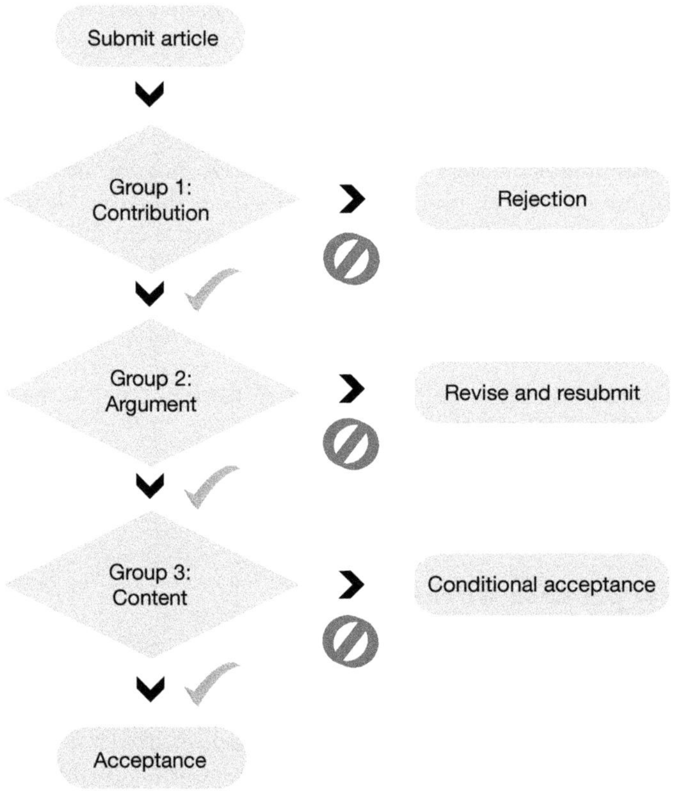

Figure 27.1 Flowchart of the editorial decision process.

> **Your research must be important, provide a novel solution, and be within the scope of the journal, or it will be rejected.**

the contribution. This is why Components 1, 2, and 3 are virtually ubiquitous in an Introduction. Perhaps the most common reason for rejection, however, is that the paper is not suitable for the journal it was submitted to, as discussed in section 27.2.

The second group of questions looks at the details of the current work. If changes are suggested at this level, it is unlikely that a judgment will be made to accept or reject the paper until those changes are made, and you will be asked to resubmit it for review after major revision. As shown in group 2 of Table 27.1, these questions focus on the overall length and organization of the paper, as well as whether you have clearly shown that you have a reasonable answer to your research question or a workable design that is better than the current standard in at least one aspect. They will also check whether your works cited are up to date, particularly whether you cite relevant recent work in

Table 27.1 Groups of questions leading to different editorial decisions

Question	*Treatment in this text*	*Location in this text*
Group 1—Negative review leads to rejection		
1 Suitability for scope of journal	Journal scope	27.2
2 Importance of problem or question	Claim 1a	4.2
3 Contribution of design or answer	Claim 1b	4.2
Group 2—Negative review leads to revise and resubmit		
1 Acceptable organization and length	Structure	3.4
2 Citations current, include journal of submission	Citations	7.2
3 Acceptable statement of methods or design	Claim 2	5.2
4 Acceptable results and conclusions	Claim 3	6.2
Group 3—Negative review leads to conditional acceptance		
1 Suitable title and abstract	Title, abstract	26.2–3
2 Tables and figures necessary, clear	Graphics	24.3–5
3 Appropriate amount of detail	Organization	21.2–3
4 Claims justified	Clarification	22.2–3
5 English usage and formatting	Revision	23.2–3

the journal you are submitting to. In science one of the most common requests at this level is for clarification of the experimental design or statistics. Specific things reviewers will look for and comment on are related to components other than importance (1) (addressed in group 1), including:[4]

> If any of the basic components are missing or poorly written, this will lead to a decision to revise and resubmit.

- The connection between past and present research is clear (2);
- The theoretical framework chosen is justified (4);
- The design process or experimental setup is described clearly (5);
- The analysis or testing method is fully documented (6);
- The summary of the results is adequate (7);
- The claims about comparisons do not exceed what is justified by the data (8);
- The explanations are based on sound logic and consider alternatives (9);
- The aims of the paper as well as the conclusion are clearly stated and match the actual design and testing (3, 10).

If the changes required are all at the level of group 3, the paper may be conditionally accepted, subject to making the required changes. This could include:

- Rewriting the title or abstract to better represent the contents of the article;
- Revision of graphics;
- Adding more data or discussion, or a single reference;

- Eliminating unnecessary detail or unsupported claims;
- Modifying the strength of claims.

If there are extensive grammar, typographical, or formatting errors that do not affect content, an editor may ask that the paper be revised by a native speaker. If so, this should be done after you have made all other required changes.

27.5 Responding to reviewer comments

As I have said many times, beginning in Chapter 1, it is difficult for a novice author to understand how many times they will have to revise their paper before it is published. Even after submission, the process is not over, as a paper is almost never accepted for publication without further revision.

Students often wonder whether it is necessary to make all the changes suggested by the editor or reviewers. It is not. There are three options for addressing reviewer comments:

- Make the suggested changes;
- Make other changes to address the concern;
- Make no change, but justify your reason.

In addition to making the changes in the document itself, you must also write a letter specifying how you addressed each of the reviewers' comments. *You must respond in some way to each comment, although it is not necessary to mention every comment individually.* It is important to be respectful in your writing, not only because it is good professional practice or because they are colleagues, but also because they will make the final decision on whether your paper is published. There is no fixed format for the response, but there are at least three general formats the response may follow, and three basic ways of ordering the individual responses.[5]

> You do not have to make all the changes suggested by the reviewers but must respond to every comment.

Just as there is a difference in the writing style of different authors, there is a difference in the style of reviewers. Some reviewers are very terse, listing suggested changes without additional explanation. Others will write extensively why they think certain changes are necessary. Depending on how long the reviewer comments are, you may choose to use one or more of the following methods when you respond:

- Highlight comments and respond;
- Extract comments and respond;
- Summarize comments and respond.

Some reviewers take time to order their comments, while others make comments as they go through the paper. Sometimes the order chosen by the two reviewers is similar, but not always. As you respond to the comments, you may choose one of several basic orders:

- Respond to each reviewer separately, in the order listed;
- Respond to both reviewers in the order changes appear in the paper;
- Respond to more important changes first, then smaller changes.

The method you choose will depend on your personal preference as well as what will make it easiest and clearest to respond to all the comments. Since most reviewers include some explanation, questions, or quotes from your writing, it is generally good to either highlight the main points you are responding to or extract them. The order chosen may depend on how much similarity there is between the reviewer comments. If there is little overlap, it may be easiest to respond to each reviewer separately. If there is a lot of overlap, it may be easiest to group similar comments and respond to them together, particularly if the two reviewers make opposing suggestions. Frequently a mixed order is used, first dealing with large changes in order of importance or appearance, with small changes such as typographical or formatting errors summarized at the end rather than listed individually.

Different formats can be used when responding to reviewer comments. As with your article, consider both what will be easy to read and also make a strong argument.

If you are making the suggested changes, less comment is required. If the error would have caused a major problem for readers, it is good to mention it separately and thank the reviewer, while typographical or formatting errors can usually be summarized, as in the following examples:

- Equation 1.3 has been corrected. We thank reviewer 1 for catching this error.
- All typographical errors noted by the reviewers have been corrected.

If you think the reviewers misunderstood what you wrote, try to think why they interpreted it that way and how you could rewrite it to make the meaning clear. This may involve rewriting it differently than suggested, including comments like the following:

- Section 2.1 has been rewritten to clarify the main point of the argument.
- This sentence has been rewritten. We thank reviewer 2 for pointing out the ambiguity.

If you feel your wording is correct and essential, you can try to justify why it is correct. For example, in an article on how to help science students learn to distinguish evidence, inference, and argumentation (Rau, 2009), I wrote the following, but the editor struck out certain words to make it more concise.

- "Thus, it is necessary ~~not only~~ to collect data, ~~but also~~ to make and test inferences and convince other scientists that your interpretation is correct."

I responded:

- "The wording is important in this sentence to emphasize that data do not speak by themselves, but are both interpreted and used as part of an argument, reinforcing the subtitle [An activity to help students learn about observation, interpretation, and argumentation]."

The editor accepted my justification and reinserted the words.

One final thing related to responding to the editor is that changes made should be highlighted to make it easier for the editor to focus on those changes. In MS Word, it is common for coauthors to make changes using the Track Changes feature. Unfortunately, many journals consider it inappropriate to send a document marked with Track Changes back to the editor. Rather, items should be changed using Track Changes, then the changes should be highlighted, and finally all changes accepted so only the highlighting shows. It is not necessary to mark small things like typographical errors.

27.6 Borrowing from your own work

A common question in academic writing is how much you can borrow from something you have written previously. It is tempting to think that, because it is your own writing, it should be acceptable to reuse sentences or even whole paragraphs elsewhere. Whether this is true or not depends largely on where the material was published.

Once something has been published in an academic journal it cannot be reused in the same form anywhere else without proper attribution, because that printed material is copyrighted and the owner of the copyright is the journal, not the author. This is why publishing the same article elsewhere, or even substantially the same information (in the same language or translation), is considered plagiarism and subject to severe penalties, including termination of employment. Short portions may generally be used if the original is cited, but larger portions can only be reproduced with permission from the publisher.[6]

Conference proceedings do not have the same weight as a journal article, so you can borrow more from them than from other written work.

Articles printed in conference proceedings are a different story. Conference papers are not considered to have the same weight as academic articles, because the review process is usually restricted to review of the abstract, not the full article. Such papers will frequently be published later as journal articles and may contain the same material and even much of the same wording, but always require rewriting based on feedback received.

A thesis or dissertation is intermediate between a journal article and a conference paper. Unlike a conference paper, it is not considered preliminary, and is considered to be published, but the author is usually the copyright holder and therefore can choose to reuse some of the material. Thus, it is expected that many of the ideas from the paper, including some of the graphics, may be reused in journal articles, but usually with extensive rewriting.

27.7 Closing thoughts

There are three basic steps to any successful writing. First, consider your goal and audience. What is your purpose, and how will this affect your claims? Who will be reading your work, why will they read it, and how will they read it? Second, look at successful exemplars. Think about the structure, the words used, the formatting. What made those exemplars successful? How do they fit the general pattern, yet stand out as excellent? Finally, begin writing. Write in stages, prewriting, then writing, then rewriting over and over until you cannot find a way to improve it—then ask someone else for comments and revise it again.

Learning to write well is a never-ending process. The more you write, the better you will get. If you take the time to think about what makes some writing more successful, you will improve faster and be better able to help others, either as a reviewer or an advisor. Enjoy the journey.

Evaluate your understanding

(1) Why might you choose to submit to different journals at different stages in your career?
(2) How does online submission differ from older methods?
(3) What factors are most likely to lead to outright rejection of a manuscript?
(4) What factors are likely to lead to a request to revise and resubmit?
(5) Do you need to make all the changes suggested by an editor?
(6) How do you write a response to the reviewer comments?

Homework

(1) Ask someone who has successfully published a paper if you could look at their response to reviewer comments. What method and order did they use in responding? Did they make all the suggested changes? If not, how did they change or justify their wording?

Hint: Some people will be more willing than others to help you in this way, as it is somewhat face-threatening. Alternatively, if you coauthor a paper with a more experienced writer, notice how they respond or modify your responses to the reviewers, and learn from that model.

Notes

1 www.ieee-pels.org/images/documents/New%20Guidelines%20for%20Manuscript%20Submission%20to%20IEEE%20TPEL_2016.pdf (downloaded 12 September 2016).
2 As graduate students, Watson and Crick published the structure of DNA in *Nature*, the most prestigious science journal in the world, but they are a rare exception.
3 One particular set of questions used for science can be found in Cargill and O'Connor (2013: 18).

4 For an excellent discussion of how to respond to comments on these areas, see Cargill and O'Connor (2013: 102–103).
5 Cargill and O'Connor (2013: 99–107) provide excellent practical suggestions and examples for responding to editors and reviewers. For a more theoretical perspective of peer review as an argumentative discussion, see Berkenkotter and Huckin (2016: 61–77).
6 Short portions can be copied under "fair use," usually fewer than 500 words of continuous text or 800 words of intermittent text, with proper citation. Reuse of figures and tables requires permission, unless they are substantially modified, in which case a citation including "adapted from . . ." is sufficient.

References

ALLEY, M. 1996. *The craft of scientific writing*, New York, Springer.

BERKENKOTTER, C. & HUCKIN, T. N. 2016. *Genre knowledge in disciplinary communication: Cognition/culture/power*, London, Routledge.

CARGILL, M. & O'CONNOR, P. 2013. *Writing scientific research articles: Strategy and steps*, New York, John Wiley & Sons.

RAU, G. 2009. A new twist on 'mystery boxes': An activity to help students learn about observation, interpretation, and argumentation. *The Science Teacher*, 76, 30–35.

Appendix 1

Generalized component list

Component claims

1 Importance (I)

a General importance
Claim: This research is important to society
Support: Often general knowledge (to those in the field)

b Specific importance
Claim: This topic is important to researchers in the field
Support: Past research (citations)

2 Need (I)

a Research gap (IMRD)
Claim: There is a gap in current knowledge or understanding
Support: Absence of past research (nothing has been done)

b Limitations (IPTC)
Claim: The current best solution is limited or less than ideal
Support: Past research (present best solution, limitations)

c Benefit
Claim: This will be beneficial to society or the field
Support: General knowledge or past research

3 Research goal (I) [main claim]

a Specific topic (goal, aim, objective, focus, question) (IMRD)
Claim: We add to previous research in this area

b Proposed solution (IPTC)
Claim: We provide a better solution to this problem

c Contribution (IPTC)
Claim: We make an important contribution in this paper

4 Framework (I or M/P)

a Accepted procedures (IMRD)
Claim: Accepted procedures or proven methods were followed
Support: Past research (citations or general knowledge)

b Problem formulation (IPTC)
Claim: The research extends or builds on an established approach
Support: Past research (advantages of previous research approach)

c Model or theoretical framework
Claim: The research is based on an accepted model or established design
Support: Past research (model, framework, mathematical basis, design)

d Justification
Claim: This is the best method to follow for our research
Support: Past research (alternative methods)

5 Research details (M/P)

a Materials obtained
Claim: Appropriate materials were obtained from reputable suppliers
Support: Names and sources of materials

b Methods followed
Claim: Care was taken to ensure good results
Support: Details of present research design and procedures

c New design
Claim: A workable new solution, product, or process was developed
Support: Details of new design, intermediate testing of parts

d New mathematical procedures
Claim: A mathematical theorem, procedure, or algorithm was developed
Support: Details of proof, lemmas, mathematical argument, algorithm

6 Testing methods (M/T)

a Physical testing
Claim: Testing was done carefully, following verifiable procedures
Support: Details of data collection method

b Computer testing (IPTC)
Claim: Testing was done on a standard data set
Support: Details of testing method

c Simulation testing (IPTC)
Claim: Testing was done under realistic simulated conditions
Support: Details of testing method

d Statistical testing (IMRD)
Claim: Testing was done using standard statistical tests and tools
Support: Names of tests and software used, details of statistical design

7 Data patterns (R/T)

a General pattern
Claim: A pattern can be discerned in the data
Support: Graphical representation or data summary

b Data transformation
Claim: The pattern is easier to see after appropriate procedures
Support: Description of data handling or transformation methods

c Exceptions to pattern
Claim: Some data do not fit the overall pattern
Support: Identification of possible exceptions, outliers, or data errors

d Classification of patterns
Claim: The patterns can be summarized into different categories
Support: Similarities in patterns

8 Comparisons (R&D/T)

a Within the article
Claim: One treatment or design is different than others
Support: Graphics, data summary

b With prediction based on hypothesis (IMRD)
Claim: Data [support/do not support] hypothesis
Support: Statistical testing

c With expectations
Claim: Data [match/differ from/exceed . . .] expectations
Support: Graphics, data summary

d With theoretical ideal (IPTC)
Claim: Results can be compared with a theoretical (mathematical) ideal
Support: Basis for ideal and comparison with it

e With previous work
Claim: Data [support/refute/improve on . . .] previous work
Support: Past research compared with present, in graphics or text

9 Interpretations (D/T)

a Analysis methods (IMRD)
Claim: Interpretation of the data was carried out in a reasonable way
Support: Details of analysis method, assumptions

b Cause of results
Claim: There is a reasonable cause for the data
Support: Explanation of data pattern based on model or design goal

c Elimination of causes (IMRD)
Claim: There is no better cause for the data
Support: Lack of support for alternative hypotheses or explanations

d Practical applications (IMRD)
Claim: Understanding gained through this research has practical use
Support: Potential applications of the new knowledge

e Limitations
Claim: Conclusions are valid within a limited range of conditions
Support: Limited range of testing or advantage

10 Conclusion (D/C)

a Summary
Claim: The research goal has been achieved
Support: Summary of research, answers to questions or new design

b Achievement or advantage (more common in IPTC)
Claim: Solution improves on past work
Support: Contribution of research

c Future work (more common in IMRD)
Claim: Remaining questions can be answered or further improvements made
Support: Planned or possible future work

Support: evidence

I Past research

a General background knowledge for those in field (often without citation)
b Definitions, terminology, or notation (with or without citation)
c Description of theoretical frameworks, models, or approaches
d Citation of specific methods, results, or conclusions from past research

II Present research

a Numbered or bulleted examples or equations
b Details and sources of materials
c Details of present research design and procedures
d Details of new design, intermediate testing of parts
e Details of proof, lemmas, mathematical argument, algorithm
f Detailed description of data collection and testing procedures
g Data summary or highlight in text
h Data patterns visualized in graphics
i Data comparison, sometimes using statistical test results

Support: reasoning

III Logical

a Deductive: Testing predictions made by a general theory
b Inductive: Compiling observations to reach a general conclusion
c Abductive: Choosing the most likely of several possible explanations

IV Evaluative

a Evaluation of previous research
b Evaluation of explanations
c Evaluation of current design

V Organization

a Initial summary to reveal argument structure of subsequent section(s)
b Reminder of the research goal
c Concluding summary of argument in preceding section(s)
d Referring to arguments made in other sections

Appendix 2

Concordance, Academic Word List, and related tools

Definitions

Concordance program: Software that produces a concordance from a corpus.

Concordance: A list showing all the words in a document or corpus, with their immediate context.

Corpus: A body of documents that are related to one another in some way.

Concordance: purpose

Concordance programs are useful for looking at words in context. For example, they could be used with the exercises in this book to:

- Identify components, based on common component markers (4.1, 5.1, 6.1)
- Identify uses of statistical terminology (8.3)
- Find citation verbs and pointing verbs (7.3, 8.5)
- Find contribution, achievement, and application markers (22.1)
- Find logical and time sequence connectors (22.2)
- Find likelihood, frequency, and measurement qualifiers, modals, delimiters, and hedges (22.3)
- See how prepositions and articles are used (23.2)

Concordance programs may be more suitable for linguists doing research than engineers and scientists learning how to write. For the former they are a tool they could continue to use, whereas the latter are unlikely to use them outside a writing class.

Concordance: steps

I will briefly introduce a family of programs developed by Lawrence Anthony, available online with excellent simple instructions. The basic concordance program is AntConc (Anthony, 2018). It and the related programs mentioned below can be downloaded free from www.laurenceanthony.net/software.html. The most current versions of the instructions are listed under "Help" for each program. Many similar programs are available elsewhere.

(1) Create a corpus.

> Science: Use AntCorGen to identify articles from the PLOS ONE database of science and medicine. Once the articles are chosen, the program will automatically generate a corpus.

Engineering: Use AntFileConverter to convert your PDF or Word files to text files. Open the files you want to convert in the online tool, click Start, and they will be converted into text files in the same directory. Note that converter programs have difficulty with equations and similar features common in engineering texts, so there will be gaps and not all lines will display properly.

(2) Create a concordance, searching for words of interest.

Use AntConc to open the text files you want to examine. Create the concordance following page 3 of the instructions. The results can be sorted alphabetically by words to the left or right of the target word. Currently the latest instructions are available at www.laurenceanthony.net/software/antconc/releases/AntConc357/help.pdf.

For example, if you want to see what prepositions are used with "correlation" in (Kleinheinz et al., 2009), the results are shown in Figure A2.1, sorted by the words following the search term.

Academic Word List

Another tool teachers may find handy is the AntQuickTools (www.laurenceanthony.net/software/antquicktools), which will highlight words from the Academic Word List (Coxhead, 2000) in any text entered, for example an exemplar article. While this will not point out specialized words specific to the discipline, students can identify those easily. Instead, it will point out more advanced general English words they may need to learn.

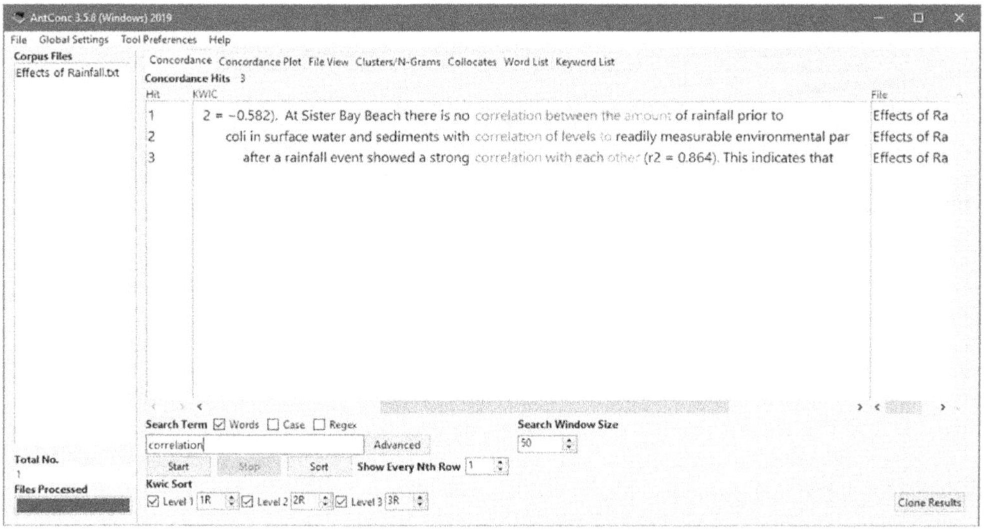

Figure A2.1 Sample AntConc output, search for "correlation" in one journal article.

References

ANTHONY, L. 2018. Software. www.laurenceanthony.net/software.html.

COXHEAD, A. 2000. A new academic word list. *TESOL Quarterly*, 34, 213–238.

KLEINHEINZ, G. T., MCDERMOTT, C. M., HUGHES, S. & BROWN, A. 2009. Effects of rainfall on *E. coli* concentrations at Door County, Wisconsin beaches. *International Journal of Microbiology*, https://doi.org/10.1155/2009/876050.

Appendix 3

List of supplemental material (online)

Teacher material

Complete contents of textbook, showing all levels of subheadings

Teaching from the textbook: schedules, assignments, and rubrics

Student material

Audio PowerPoint introductions of each chapter (about 15 minutes each)

Chapters 4–6: sample component analysis of two articles

Chapter 22: supplemental material (particularly useful to ELF students)

Class exercises

9.1 Sequencing strategies

19.1 Linearizing

21.1 Coherence (with answer key)

21.2 Coherence 2

21.3 Conciseness (with answer key)

21.4 Conciseness 2

22.1 Logical connectors (with answer key)

23.1 Eliminating ambiguity (with answer key)

23.2 Articles (with answer key)

24.1 Evaluating graphics

25.1 Reference formats

Glossary

7Cs of Change A stepwise rewriting process including coherence, conciseness, connection, connotation, consistency, correctness, and collaboration.

Argument A claim supported by evidence and reasoning.

Argument structure A summary of the claims, component claims, and types of support used in a document, related to the division structure.

Author-prominent citation Citation of previous work in which the author's name appears as part of the sentence, also called integral citation.

Citation verb Verb referring to what a previous author did.

Citation Mark in the text indicating the source of information used to support a statement. See author-prominent citation and information-prominent citation.

Claim An assertion or proposition, which may include one or more component claims, requiring support.

Component Component claim, a subclaim commonly found in science and engineering research writing, often in a predictable order, with or without recursive cycles.

Component analysis Identification of the components and support present in an exemplar.

Component markers Common phrases or groups of words without specific content that help readers identify components and that can be reused without fear of plagiarism.

Condensed style Writing style that assumes the reader has specialized knowledge of the subject, often with many ideas in one sentence.

Division Conceptual units of a research article (either IMRD or IPTC) that can be distinguished by their basic claims, and may include one or more sections.

English as a lingua franca (ELF) The use of English as a common language of communication among speakers of various first languages.

Evidence Details about or data from the present work or citations of previous work, used to give an indication whether a claim is true or false.

Exemplar A representative work of a particular genre that can be used as an example of the style expected in that genre.

Experimental design Details of the subjects, variables, measurement, and testing methods used in quantitative research.

Extended style Writing style that is easy for a reader to follow, often with an initial summary followed by details and one main idea per paragraph.

Framework The theoretical framework, model, or accepted procedures followed or modified in the present research.

Genre A distinct style of writing or speaking. Defined by linguists as a class of communicative events within a discourse community to accomplish a shared set of communicative purposes.

Graphics All the tables and figures in a document. Figures include graphs and illustrations.

IMRD Introduction, (Materials and) Methods, Results, Discussion. The four divisions of a prototypical science research article.

IMRaDC Introduction, (Materials and) Methods, Results and Discussion, Conclusion. A modification of IMRD. Common in some sciences and branches of engineering dealing with more quantitative research, such as chemical engineering and some fields of civil engineering.

Inference A logical conclusion, either inductive or deductive, based on the evidence, often incorporating background knowledge or assumptions.

Information-prominent citation Citation of previous work in which the author's name does not appear in the sentence, also called non-integral citation.

IPC Introduction, Proof, Conclusion. A modification of IPTC. Common in mathematics, theoretical physics, and some fields of electrical engineering that focus on mathematical proof and therefore do not require empirical testing.

IPTC Introduction, (Product or) Process, Testing, Conclusion. The four divisions of a prototypical engineering research article. Also used in some branches of physics dealing with design or theory rather than quantitative research.

Plagiarism Using ideas and words of another person as if they were your own, or using copyrighted material without proper citation or permission.

Pointing verb Verb used to point a reader to graphics or to information found in other sections of the article.

Prewriting Gathering (brainstorming) and ordering (linearizing) of ideas before writing.

Recursive Occurring in cycles, as when a series of components is repeated, e.g. repeating Components 6, 7, 8 in the Testing division to test and compare different parts of a new design.

References Listing of complete bibliographic information of all works cited in the document, usually found at the end without a section number.

Research article A report of research conducted by one or more authors, published in a refereed journal, also called article, journal article, or academic paper.

Research gap Identification of an important question or problem not sufficiently covered by previous research, associated with Component 2: need for current work.

Research goal Identification of the goal of the present research, whether expressed as research question, aim, objective, purpose, focus, or problem definition.

Section Portion of a research article designated by a section number (usually) and title, often equivalent to the divisions in IMRD.

Significant In science, observations that are unlikely to occur by chance, as determined by statistical testing, thus likely indicative of a causal relationship.

Statistics Use of mathematical inference to generalize results from a sample to a larger population. Common in science but not in engineering.

Style See extended style, condensed style.

Support Evidence and reasoning used to show the veracity of a claim.

Index

For Product Safety Concerns and Information please contact our EU
representative GPSR@taylorandfrancis.com
Taylor & Francis Verlag GmbH, Kaufingerstraße 24, 80331 München, Germany

www.ingramcontent.com/pod-product-compliance
Ingram Content Group UK Ltd.
Pitfield, Milton Keynes, MK11 3LW, UK
UKHW051943061126
477531UK00005B/228

* 9 7 8 1 1 3 8 3 8 8 2 4 6 *